On the Modern Cult of the Factish Gods

Science and Cultural Theory

A Series Edited by
Barbara Herrnstein Smith
and
E. Roy Weintraub

On the Modern Cult
of the Factish Gods

Bruno Latour

First chapter translated by
Catherine Porter
and
Heather MacLean

Duke University Press
Durham and London
2010

Printed in the United States of America on acid-free paper ∞
Typeset in Quadraat by Tseng Information Systems, Inc.
Library of Congress Cataloging-in-Publication Data
appear on the last printed page of this book.

Duke University Press gratefully acknowledges the support
of the French Ministry of Culture, which provided funds towards the
translation of "On the Modern Cult of the Factish Gods."

Contents

LE STATUAIRE ET LA STATUE DE JUPITER .Fable CLXXV.

FIGURE 1.

J.B. Oudry illustration for the Fable CLXVV, *The Sculptor and the Statue of Jupiter*, in *Fables choisies*, vol. 3, illustrated edition (Paris, 1755), 117.

Preface

Let us start with this engraving by Jean-Baptiste Oudry, entitled *The Sculptor and the Statue of Jupiter*, from the beautiful edition of the *Fables* published in 1755 (fig. 1). It is most likely that the artist, overtaken by the fable, has somewhat exaggerated La Fontaine's intent: in the morning the sculptor enters the workshop, where the night before he had given the last touch to his statue, and he now stands with arms extended, terrified, expecting Jupiter himself to strike him to dust with his thunderbolts.

> Indeed, the man whose skill did make
> Had scarcely laid his chisel down,
> Before himself began to quake,
> And fear his manufacture's frown.

Have you ever met an artist so naive that he could be thus impressed? The fabulist seems to believe it since he claims that such a naiveté is precisely the origin of the sin of idolatry:

> This trait we see in infancy
> Between the baby and its doll,
> Of wax or china, it may be—
> A pocket stuffed, or folded shawl.
> Imagination rules the heart:
> And here we find the fountain head
> From whence the pagan errors start,
> That over the teeming nations spread.
> With violent and flaming zeal,
> Each takes his own chimera's part;
> Pygmalion does a passion feel
> For Venus chiseled by his art.

Is this not well known? Pagans are like kids who let themselves be taken in by their own chimeras: at night they fabricate statues, poems, dolls and myths; in the morning they believe that these have made themselves through spontaneous generation, and that they deserve a cult or to be loved madly. None of these fabricators understands that they have manufactured their objects: the sculptor is terrified by his statues, the child by her dolls, the poet by the gods he has invented. As to Pygmalion, he is not only stupid but also incestuous, since he fell in love with Venus, his own daughter in stone. All offer cases of this disease of the mind, which has since been called "fetishism," and by which the fabricators fall into the trap of their own contraptions. La Fontaine concludes:

> All men, as far as in them lies,
> Create realities of dreams.
> To truth our nature proves but ice;
> To falsehood, fire it seems.

This is a surprising moral coming from the pen of a fabulist! Is it true that one should remain so icy that no reality whatsoever should be given to one's dreams? Such is an implausible portrait of a rational being: what? A poet without poems, a sculptor without statues, a child without dolls, a fabulist without fables; a rational and naked human, obliged, in order to regain a normal temperature, to annihilate the work of his own hands? Here, an even more unlikely portrait of an irrational human being: before the artist has been taken in by his own creation, he benefited from a total freedom, as we learn at the beginning of the poem:

> A block of marble was so fine,
> To buy it did a sculptor hasten.
> "What shall my chisel, now It's mine—
> A god, a table, or a basin?"
> "A god," said he, "the thing shall be;
> I'll arm it, too, with thunder.
> Let people quake, and bow the knee
> With reverential wonder."

What surprising changes of temperature from ice to fire: do you know any artist so bipolar that she abruptly alternates from a feeling of total freedom to that of total servitude? Is it the way we all create? Is it the way we have been created? Is it the sort of life we breathe into our own beings? Either you have to destroy all the things you have created in order to remain cool as a rational being, or else you will be taken in by the creatures of your own hands: as if there were no plausible passage between fetishism and iconoclasm.

La Fontaine is making fun of everybody, no doubt about that: of the idolaters yes, but also of those who claim to have no illusion and who believe the idolaters to be naive, and thus also of himself. Oudry, the remarkable engraver, is also mocking himself, La Fontaine, the sculptors, and the gods. In his engraving it is Jupiter who opens his arms, in astonishment and terror at the view of his maker, who suddenly opens the workshop's door, as if the god had been taken by surprise! Everything is wrong in this fable; everything is right; everything is to be taken up again. It is not a fable for nothing!

Because this engraving illustrates so well the double bind of creation, it could serve as the emblem of this book. On the one hand, it tells us that we have to choose between fiery illusions and icy reasons, but on the other, it tells us the opposite, since there is no way to choose between the two, as soon as we put our hand into the creation of any thing. This is a double contradiction, one explicit, the other implicit and folded into the work of art itself.

The Moderns possess this interesting peculiarity of always dreaming of a temperature that they never had any chance of attaining. In order to draw their anthropological portrait, it is necessary, as La Fontaine indicates, to decrypt their fable and detect the secret passage they have unwittingly dug between fetishism and iconoclasm.

To inquire into this double bind I propose here two makeshift notions: that of *factish* and that of *iconoclash*. I might be pardoned those two neologisms, if I say that they are the result of two rather distinct moments of fieldwork. The first is the outcome of an internship that I did in 1995, in the consultation of the ethnopsychiatry of Tobie Nathan, at the Centre

Devereux in Paris. I wanted to check the vast literature on fetishism and anti-fetishism against the actual practice of a real practitioner. The first chapter of the present book is the translation of the pamphlet I published on this experience.[1] Then, for four years, I had the good fortune of preparing the exhibition *Iconoclash* for which I was curator, with several friends, and which opened in 2002. If the notion of *factish* allows one to suspend the belief of belief, that of *iconoclash* aims at suspending iconoclastic gestures. Instead of another iconoclastic exhibition, we wanted to do an exhibition *about* iconoclasm, transforming this notion from a resource to a topic. The sumptuous catalog being out of print, it seemed a good idea to republish its introduction.[2] Since the question of the right way to consider images and more generally mediations is at the core of all the issues of fetishism and iconoclasm, I have added a third chapter in which I compare religious images to those of scientific practice, so as to propose another take on the science and religion debate.[3]

The book thus composed requires nothing more from the reader than the suspension, no doubt momentary, of beliefs in belief and in critique. It is, in my experience, the only way to focus on the creations of our hands, and to pay one's exact dues to creator and to creatures alike.

CHAPTER ONE

On the Cult of the Factish Gods

"The light-skinned peoples living in the northern reaches of the Atlantic are said to have a peculiar way of worshipping the gods. They go on expeditions to other nations, seize statues of their gods, and destroy them in huge bonfires, insulting them with cries of 'Fetish! Fetish!'—a word that in their barbaric language seems to mean 'forgery, nonsense, lie.' Though they insist that they have no fetishes, and that it was their own idea to free other nations from such things, they seem to have very powerful gods. Indeed, their expeditions frighten and fill with dread the peoples who are attacked in this way by rival gods, who these peoples call 'Moh Dun,' and whose power appears as mysterious as it is invincible. It seems that in their own lands they have built many temples, and the way they worship inside them is as strange, frightening, and barbaric as it is outside. During great ceremonies repeated from generation to generation, they smash their idols to pieces with hammers. They seem to benefit significantly from these ceremonies, for once they have freed themselves from their gods they can do whatever they please. They can mingle the forces of the Four Elements with those of the Six Kingdoms and the Thirty-Six Hells, without feeling at all responsible for the violence they unleash. Once these orgies have ended, these people are said to fall into deep despair. At the feet of their shattered statues they cannot help but hold themselves responsible for everything that happens, which they call 'human' or 'free-will subject'—or else they believe, on the contrary, that they are responsible for nothing at all, and that they are entirely produced by what they call 'nature' or 'causal objects' (the terms are hard to translate into our language). Then, as if terrified by their own daring, and in order to put an end to their despair, they repair the Moh Dun gods they have just broken, making countless offerings and sacrifices; they put their gods back up at the crossroads, holding them together by iron hooping as we do for barrel staves. They are also said to have created a god in their own image—in other words, one just like themselves, sometimes absolute master of all he does, and sometimes completely nonexistent. These barbaric peoples do not seem to understand what it means 'to act.'"

—Reported by Counselor De-Bru-Osh, emissary to China from the
Korean Royal Court in the mid-eighteenth century

Part One
Fairy-Objects, Fact-Objects

In order to mock both our own foolish beliefs and those of others, our freethinking ancestors handed down to us the scoffing tone set by Voltaire, among others. But in order to mock all forms of worship in this way, in order to overthrow all our idols, we had to believe in reason, the only force capable of seeing through all those follies. How can we speak symmetrically about ourselves and others without believing in either reason or belief, while still respecting both fetishes and facts? I have tried my hand at this, somewhat awkwardly, by defining *agnosticism* as a way of ceasing to believe in belief.

How the Moderns Fabricate Fetishes in the Lands of People They Meet

Belief is not a state of mind but a result of relationships among peoples; we have known this since Montaigne. The visitor knows; the person visited believes. Conversely, the visitor knew, the person visited makes him understand that he only thought he knew. Let us apply this principle to the case of the Moderns. Wherever they drop anchor, they soon set up fetishes: that is, they see all the peoples they encounter as worshippers of meaningless objects. Since the Moderns naturally have to come up with an explanation for the strangeness of a form of worship that cannot be justified objectively, they attribute to the savages a mental state that has internal rather than external references. As the wave of colonization advances, the world fills up with believers. A Modern is someone who believes that others believe. An agnostic, conversely, does not wonder whether it is necessary to believe or not, but why the Moderns so desperately need belief in order to strike up a relationship with others.

It all started on the West Coast of Africa, somewhere in Guinea, with the Portuguese. Covered with amulets of the saints and the Virgin themselves, they accused the Gold Coast Blacks of worshipping fetishes. When the Portuguese demanded an answer to their first question, "Have you made these stone, clay, and wood idols you honor with your own hands?"

the Guineans replied at once that indeed they had. Ordered to answer the second question, "Are these stone, clay, and wood idols true divinities?" the Blacks answered "Yes!" with utmost innocence: yes, of course, otherwise they would not have made them with their own hands! The Portuguese, shocked but scrupulous, not wanting to condemn without proof, gave the Africans one last chance: "You can't say both that you've made your own fetishes and that they are true divinities; *you have to choose*: it's either one or the other. Unless," they went on indignantly, "you really have no brains, and you're as oblivious to the principle of contradiction as you are to the sin of idolatry." Stunned silence from the Blacks, who failed to see any contradiction, proved how many rungs separated them from full and complete humanity. Bombarded with questions, they persisted in repeating that they did make their own idols, and *therefore* these were indeed true divinities. Confronted with such blatant bad faith, the Portuguese could only respond with jeers, derision, and disgust.

To designate the aberration of the coastal Guinea Blacks, and to cover up their own misunderstanding, the Portuguese (very Catholic, explorers, conquerors, and to a certain extent slave traders as well) are thought to have used the adjective *feitiço*, from *feito*, the past participle of the Portuguese verb "to do, to make." As a noun, it means form, figure, configuration, but as an adjective, artificial, fabricated, factitious and finally, enchanted.[1] Right from the start, the word's etymology refused, like the Blacks, to choose between what is shaped by work and what is artificial; this refusal, this hesitation, induced fascination and brought on spells. Even though all etymological dictionaries agree on the origins of the term, Charles de Brosses, who invented the word "fetishism" (French fétichisme) in 1760, linked its origins with fatum, or destiny, the source of the French noun fée, "fairy," and of the adjective form in the noun phrase objet-fée, "fairy-object" (also of the English adjective "fey").

"The Blacks from the west coast of Africa, and even those from the interior, all the way up to Nubia, along the Egyptian border, worship certain divinities that the Europeans call fetishes, a term de-

vised by our traders from Senegal based on the Portuguese word
Fetisso [sic], that is to say, a fairy-object, an enchanted, divine, or
oracular object, from the Latin root Fatum, Fanum, Fari."[2]

Whatever root we may prefer, the either-or choice remains the one on
which the Portuguese insisted and the Blacks rejected:

> "Who is speaking in the oracle? Is it the human being, or the
> fairy-object itself? Is the divinity real or artificial?"
>
> "Both," the defendants reply at once, since they are unable to
> grasp the difference.
>
> "You have to choose," say the conquerors, without further hesi-
> tation.

The two roots of the word indicate rather well the ambiguity surround-
ing an object that talks, that is fabricated or, to blend both meanings
into a single expression, an object that provokes talk. Yes, the fetish is a
"talk-maker."

Too bad the Africans did not return the compliment. It would have
been nice to see them ask the Portuguese dealers if their own amulets of
the Virgin had been made by hand or had fallen directly out of the sky.

> "Carefully crafted and engraved by our goldsmiths," they would
> have answered proudly.
>
> "So are they really sacred?" the Blacks would have asked.
>
> "Of course they are; they were solemnly blessed by the arch-
> bishop in Nossa Senhora dos Remédios church, in the presence of
> the King."
>
> "So if you recognize both the working of gold and silver in the
> smith's crucible and the sacred character of your icons, why are you
> accusing us of contradiction, when we're saying the same thing?
> What's good for the goose is good for the gander."
>
> "Sacrilege! No one can confuse idols to be smashed with icons
> to be prayed to," the Portuguese would have answered, indignant
> all over again in the face of such impudence.

Still, we can bet that they would have called upon a theologian to get them out of the predicament into which they had been thrown by the merest hint of symmetrical anthropology. But they would have needed a subtle scholar to teach them how to distinguish between *latria* (excessive adoration, reserved for God) and *dulia* (moderate adoration as for example of the Virgin Mary). "Pious images," the theologian would have intoned, "are nothing in and of themselves; they simply serve to remind us of the model that is the only legitimate object of worship. Your monstrous idols, on the other hand, are supposed to be the divinities themselves, from what you say, and yet you impudently admit that you've made them yourselves from scratch." Why should he jeopardize his reputation in a theological discussion with mere natives, anyway? Ashamed of equivocating, in the grip of a holy zeal, the theologian would have toppled the idols, burned the fetishes, and then consecrated the True Image of the suffering Christ and his Holy Mother inside the disinfected temples.

Even without the help of this imaginary dialogue, we can see perfectly well that what we have here is not a contrast between idolatrous Gold Coast Blacks and image-free Portuguese visitors. We see one group of people covered with amulets scoffing at another group of people covered with amulets. We do not have iconophiles on one side and iconoclasts on the other, but iconodules on both sides (one side being made of selective iconoclasts). Yet the misunderstanding persisted, because each side, acting on its own terms, refused to choose. The Portuguese refused to hesitate between true objects of piety and sinister masks covered with sacrificial blood and grease. On the Gold Coast, every Portuguese suddenly displayed the fervent indignation Moses expressed against the Golden Calf. "Idols have mouths, but never speak, eyes, but never see, ears, but never hear." The Guineans, on their side, could not see any obvious difference between the idol that had been brought down and the icon erected in its place. Relativists before the term was invented, they thought what the Portuguese were doing was the same thing they did. And it was precisely this failure to distinguish, this lack of comprehension, that condemned them in the eyes of the Portuguese. These savages could not even tell the difference between *latry* and *dulia*, between their own fetishes and the holy icons of the invaders; they refused to grasp the extent of the gulf that separated human construction of an artifact from the definitive reality of what no human has ever constructed. Even the difference between immanence and transcendence seemed to be beyond them. How could the Portuguese fail to see them as primitives, and fetishism as a primitive religion?[3]

All the more so since the savages have been diabolically persistent in their error. Three centuries later in contemporary Rio de Janeiro, Black and Portuguese mestizos stubbornly maintain both that their divinities are made, fabricated, "seated," and that, as a result, they are real. The anthropologist Patricia de Aquino has collected and translated accounts by Candomblé initiates:

> "I was shaved (initiated) in Salvador for Osala, but I had to seat Yewa (who asked, through divination, to be seated, installed, made, fabricated), and Mother Aninha (his initiator) sent me to Rio because at the time Yewa was already an endangered Orisa, so to speak. There were many who no longer knew the oro (Yoruba term for "the words and rites") of Yewa."

> "I am from Oba, Oba is almost dead already because no one knows how to seat her, no one knows the craft, so I came here (to this Candomblé) because I was shaved here, and they are not going to forget the awo (Yoruba term for "secrets") for making her."[4]

The anti-fetishist slumbering within us cannot stand the brazenness of such statements. Hide the construction process, the craft, the *fazer* that we cannot see! How can you so sanctimoniously admit that you have to make, fabricate, seat, situate, construct these divinities that grip you and yet remain out of reach? Are you so unaware of the difference between making what comes from yourselves and receiving what comes from elsewhere?

No matter where they landed, the Portuguese, struck by the same sort of impudence, had to understand fetishism by likening it either to naiveté or cynicism. If you admit that you fabricate your own fetishes yourselves, you must then acknowledge that you pull their strings as a puppeteer would. The whole thing is engineered to impress others through disguising itself. Manipulators of popular beliefs, you then join the whole crowd of priests and falsifiers who—according to the anti-clericals—make up the long history of religions. Or else, if you let your own marionettes take you by surprise and you start believing in the airs they (or, rather, you) put on, this proves such a degree of naiveté that you are condemned to join the eternally credulous and hoodwinked masses who make up—again according to lucid observers—the gullible rabble of the history of religions.[5]

From the mouths of the Fontenelles, the Voltaires, and Feuerbachs of the world, the same either-or alternative keeps spewing forth: "Either you are cynically pulling the strings, or else you are being had." Or, even more naively: "Either you built it, or else it's real."[6] And the shaven adepts of the Candomblé persist gently: "I am from Dada, but since no one knows how to fabricate Dada, we give to Sango or Osala so that they will take over the person's head."[7] Whereas the initiates are designating something that is neither completely autonomous nor completely constructed, the notion of belief splits apart their delicate operation, the fragile bridge connecting fetish and fact, and allows the Moderns to see all other peoples as naive believers, skillful manipulators, or self-deluding cynics. Yes, the Moderns refuse to listen to the idols; they split them apart like coconuts, and from each half they take two forms of dupery: you can deceive others, and you can deceive yourself. Moderns believe in belief in order to understand others; initiates do not believe in belief either to understand others or to understand themselves. Can we recover their way of thinking for our own use?

How the Moderns Manage to Build Fetishes in Their Own Lands

If we allow ourselves to learn from those who do not believe in belief, we notice that the Moderns do not believe in it any more than the coastal Blacks did. The Whites may have accused the savages of fetishism, but this did not necessarily make them naive anti-fetishists themselves. If we believed that, it would move us further from Scylla but closer to Charybdis; if we believed that, we would save the Blacks from belief—and belief would then become an accusation leveled by the Whites against something they did not understand—but we would be plunging the Whites into an abyss of naiveté. We would have them believing that the others believe! We would be confusing Whites with Blacks! What we have just done for the fetishists we must do now for the anti-fetishists, and show ourselves to be as charitable toward the latter as we have been toward the former.

As it happens, just as the charge of fetishism completely fails to describe the practices of the coastal Blacks, the claim of anti-fetishism likewise fails to explain the practices of the Whites. Wherever they install their great fetish-smashing machines, the Whites begin once again to produce the same sort of uncertain beings the Blacks produced, and it is impossible to tell whether these beings are constructed or collected, immanent or transcendent.[8] Let us consider, for example, everything the fetish object is capable of doing, even though it is accused of doing nothing.

What is the definition of an anti-fetishist? An anti-fetishist is someone who accuses someone else of being a fetishist. And what is the content of the complaint? The fetishist is accused of being mistaken about the origin of the power in question. He has built an idol with his own hands—his own human labor, his own human fantasies, his own human powers—yet he attributes this labor, these fantasies, and these powers to the very object that he has created. The fetish—at least according to the anti-fetishist—acts, so to speak, like an overhead projector. The image comes from the professor who has placed a transparency on the glass over the blinding light, but what is shown seems to spring from the screen toward the audience, as if neither the professor nor the overhead projector had anything to do with it. The fascinated spectators "attribute an autonomy to the image" that it does not possess. Overturning fetishism thus amounts to inverting an inversion, reversing a reversal, rectifying the image, and granting the real master of the action credit for initiating it. Along the way, however, the real master has disappeared! The object, which was nothing, is now doing something. As for the origin of the action, it gets lost in a frightfully mixed-up battle of inheritance.

As soon as the anti-fetishist unveils the idol's ineffectiveness, he in fact plunges into a contradiction from which he cannot escape. Just when the fetish is deemed to be nothing at all, it begins to act and shift everything about. It is capable, in particular, of reversing the origin of power. Better yet, since according to the anti-fetishists the effect of the fetish is efficacious only if its creator is unaware of its origin, it must be capable of completely dissimulating its own manufacture. Thanks to the fetish, in a single wave of a magic wand, its creator can turn himself from a cynical manipulator into an ingenuous dupe. Thus, even though the fetish is nothing but what a human makes of it, it nevertheless adds a little something: it inverts the origin of the action, it dissimulates the human work of manipulation, and it transforms a creator into a creature. How could anyone deny the efficaciousness of an object that is capable of so many prodigious feats?

But the fetish does better still: it modifies the very quality of human action and work, and yet, by revealing that only human action gives voice and power to objects, the critical thinker ought to invert the inverted origins of power, ending the illusion of the fetishes once and for all. Someone who (naively) believed he was hearing voices would then turn into a ventriloquist. Having become aware of his own double-dealing, he would be reconciled with himself. Someone who believed he was dependent on divinities would notice that he is actually alone with his own inner voice, and that divinities own nothing he has not given them. Once the scales had fallen from his eyes, he would see that there was nothing to see. He would have ended his alienation—mental, religious, economic, and political—since no alien would ever again come to latch on, parasitically, to something he had built with his own calloused hands and his own creative spirit. Carried away by the critic's denunciation, humans would finally realize that they are sole masters in a world forever emptied of its idols. The fire that Prometheus had stolen from the gods would be stolen back from Prometheus himself by critical thinking. Fire would come from humans, and from humans alone.

Alone? Not quite, and this is where things become complicated once again. Like an earnest lawyer who has to divide an estate when there is no will and no heir, the critical thinker never knows to whom he should restore the power that was mistakenly attributed to the fetishes. Should it be given back to the individual who is master of himself as well as of the Universe, or to a society of individuals? If the answer is that one should render unto society that which is society's, mastery is lost all over again. The inheritance recovered from the fetishes is dispersed among a cloud of legitimate heirs. Once idolatry's reversal has been reversed, once the projection of power has been "projected" back onto an overhead, it is not the "I," the working individual, who is found at the end of the road, but a group, a multitude, a collective. Under the now-dispelled fantasy of the fetish, the enlightened human being realizes that he is not really alone, but that he shares his existence with a crowd of actors. The alien he thought he was eliminating comes back in the frightfully complicated form of a social multitude. The human actor has merely exchanged one form of transcendence for another. We can see this quite well in Emile Durkheim, in whose hands that which is social seems hardly less opaque than the offending religion it explains.[9] Marx, in his famous definition of the fetishism of commodities, illustrates how something that does nothing can still manage to proliferate:

There . . . is a definite social relation between men, that assumes, in their eyes, the fantastic form of a relation between things. In order, therefore, to find an analogy, we must have recourse to the mist-enveloped regions of the religious world. In that world the productions of the human brain appear as independent beings endowed with life, and entering into relation both with one another and the human race. So it is in the world of commodities with the products of men's hands. This I call the Fetishism which attaches itself to the products of labour, so soon as they are produced as commodities, and which is therefore inseparable from the production of commodities.[10]

Economic anthropology attests to the fact that relationships among human beings, whether fetishized by way of merchandise or not, seem no more simple or transparent than those among divinities.[11] If mer-chandise loses its seeming autonomy, no human regains mastery as a result, and certainly not the tireless worker. Through some strange in-version of the inversion, it appears that *the fetish-less world is populated by as many aliens as the world of the fetish.* The inversion of the inversion leads to a universe that is as unstable as the world that was supposedly inverted by an illusory belief in fetishes. Neither anti-fetishists nor fetishists know who acts and who is mistaken about the origins of action, who is master and who is alienated or possessed. Thus, even among the Moderns the fetish, far from being drained of its efficacy, always seems to act in such a way as to shift, muddle, invert, and perturb the origins of belief, as well as the very certainty that mastery is possible. The fetish immediately re-gains the power that people seek to deny it. No one believes. The Whites are no more anti-fetishist than the Blacks are fetishist. It is just that the Whites always erect idols in other peoples' lands and then immediately overthrow them, multiplying the operators that disseminate the origins of the action in their own lands. Yes, the anti-fetishists are just like the fetishists: they worship idols in a rather strange manner, one that we are going to have to untangle.[12]

How the Moderns Struggle — and Fail — to Distinguish Facts from Fetishes

Why must the Moderns resort to complicated forms in order to be-lieve in others' naive beliefs, or in knowledge without belief among themselves? Why must they act as if others believe in fetishes, while they seemingly practice the most austere anti-fetishism? Why not just admit that there is no such thing as fetishism — and no anti-fetishism either — and recognize the strange efficacy of these "action displacers" with which our lives are intimately bound up?[13] The reason is that Mod-erns are strongly attached to the conviction that there is an essential difference between facts and fetishes. The goal of belief is neither to explain the mental state of fetishists nor to account for the naiveté of anti-fetishists. Belief depends on something completely different: on the distinction between knowledge and illusion, or rather, as we shall see in the following sections, on the separation between practical life —

which does not make this distinction — and theoretical life, which maintains it.

Let us look a little more closely at the double repertoire that the notion of belief has to contain in separate compartments, and see how it works. As soon as the anti-fetishist has denounced naive belief and revealed the work of the human actor, work that is mistakenly projected onto idols made of stone and wood, he goes on to denounce the naive belief that the individual human actor thinks he can attribute to his own actions. It is not easy, in the eyes of the anti-fetishists, to behave like ordinary actors! If you try to dance to their tune, you are always starting off on the wrong foot. If you believe you are being manipulated by idols, they will show you that you have created them with your own hands; but if you proudly boast of your ability to create so freely, they will show you that invisible forces are manipulating you and making you their agent without your knowledge. The critical thinker triumphs twice over the consummate naiveté of the ordinary actor, seeing the invisible work that the actor is projecting onto the divinities who manipulate him, but also seeing the invisible forces that drive the actor, who believes he manipulates freely! (Critical thinkers, offspring of the Enlightenment, ceaselessly manipulate invisible things themselves, as we can see; the great liberators from alienation produce endless numbers of aliens.)

How do the Moderns go about framing the actions of ordinary actors by means of two such contradictory denunciations? They do this by using two operators instead of just one: they invoke fairy-objects on the one hand, and fact-objects on the other. When they denounce the naive belief of actors in fetishes, they are using human action that is free and focused on the subject. But when they denounce the naive belief of actors in their own subjective freedom, the critical thinkers are using objects — as they are known by the objective sciences — that they have established, and in which they place their full trust. They thus alternate between fairy-objects and fact-objects, so they can show off to ordinary, naive people twice.

Since the situation is likely to become complicated, perhaps a diagram will help guide us. Let us start with the first critical denunciation. The human actor thinks he is determined by the power of objects, a power that tells him how to behave. Fortunately, the critical thinker is watching out for him, and denounces the actor's double-dealing, which, "in reality," projects the power of his own action onto an inert object.[14]

One might believe that the work of denunciation is over. Sobered up, freed, de-alienated, the subject takes back the energy that used to belong to him and refuses to grant his imaginary constructions an autonomy that they can never again recapture. The work of denunciation does not stop here, however; it starts up again, but now in the other direction. The free and autonomous human subject boasts, a little too soon, that he is the primal cause of all his own projections and manipulations. Fortunately, the critical thinker, who never sleeps, once again reveals how determination works, beneath the illusion of freedom. The subject believes that he is free, while "in reality" he is wholly controlled. In order to explain the determinations involved, we must take recourse to objective facts, revealed to us by the natural, human, or social sciences. The laws of biology, genetics, economics, society, and language are going to put the speaking object, who believed himself to be master of his own deeds and acts, in his place.

The two forms of denunciation look strikingly similar: the critical thinker with his belief in causes (fig. 2) occupies the same position as the naive individual with his belief in idols (fig. 3). If anything appears

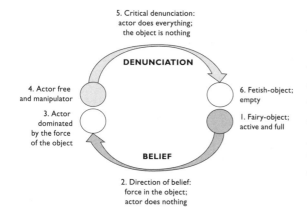

5. Critical denunciation: actor does everything; the object is nothing

DENUNCIATION

4. Actor free and manipulator

3. Actor dominated by the force of the object

6. Fetish-object; empty

1. Fairy-object; active and full

BELIEF

2. Direction of belief: force in the object; actor does nothing

FIGURE 2. The first critical denunciation reverses the direction of belief (from 1 to 3), so as to foreground the role of the actor freely projecting values upon mere objects (from 4 to 6).

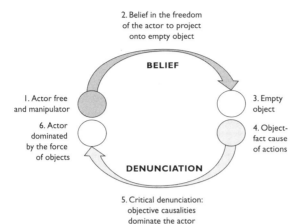

2. Belief in the freedom
of the actor to project
onto empty object

BELIEF

1. Actor free
and manipulator

3. Empty
object

6. Actor
dominated
by the force
of objects

4. Object-
fact cause
of actions

DENUNCIATION

5. Critical denunciation:
objective causalities
dominate the actor

FIGURE 3.

The second critical denunciation reverses once again the direction of belief (from 1 to 3), so as to foreground causal factors that strictly limit the actor's freedom (from 4 to 6).

to be denounced by the superimposition of the two diagrams, it ought to be denunciation itself, since this is what once again reverses the origin of the power whose reversed origin it had previously reversed! But denunciation by critical thinkers is no more at stake than is the naive belief of ordinary actors. The notion of belief allows the Moderns to understand the origin of action in their own way, through the double vocabulary of fetishes and facts.

The two diagrams presented above can never be superimposed, however, and the task of belief is precisely to prevent such superimposition. Why? Because critical denunciation is based on four different lists, two for the object pole, and two for the subject pole—four lists that must never be mixed together under any circumstances. To state it bluntly, the critical thinker will put everything he does not believe in on the list of fairy-objects—religion, of course, but also popular culture, fashion, superstitions, mass media, ideology, and so on—and he will put everything in which he firmly believes on the list of cause-objects: economics; sociology; linguistics; genetics; geography; neuroscience; mechanics; and so on. Conversely, he will constitute his subject pole by putting all the aspects of the subject that are dear to him—responsibility, liberty, inventiveness, intentionality, and so forth—on the credit side of the ledger, and will put on the debit side anything that he deems useless or artificial: mental states; affects; behaviors; fantasies; and the like. The content and length of each list will vary from thinker to thinker, but the four-part division will remain intact.

Thus, as the anti-fetishists see it, naive belief starts off on the wrong foot every time. It credits fetish-objects with a power that comes from human ingenuity alone; this is what the first denunciation makes brutally clear (fig. 4). Naive belief also credits itself with freedom when it is actually manipulated by a host of causal determinations; this is what the second critical denunciation obligingly reveals. But the resemblance between the two ways of proceeding never strikes the anti-fetishist, because the fact-object used in the second critique comes from a list of solid objective causes, while the fairy-object denounced in the first critique is only the projection of a hodge-podge of more or less vague beliefs, heaped onto an unimportant substratum. Conversely, the active subject carrying out the first denunciation is assigned the role of a human actor rebelling against alienation and courageously demanding his full and total freedom, while the active subject in the second denunciation is a puppet drawn and quartered by all the causal determinations that mechanize it in all directions. Provided that the black and grey lines

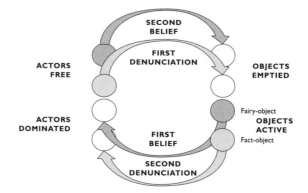

FIGURE 4.

The two critical denunciations might appear contradictory, except if a strict division is enforced between the two types of objects (on the right), and the two types of subjects (on the left) that both critiques presuppose. It is this strict division that, taken together, constitutes the phenomenon of belief.

of figure 4 are kept strictly apart, critical thought will have no difficulty claiming both that a free and autonomous human actor creates his own fetishes, and that the objective determinations revealed by the physical or social sciences define that actor completely.

We can now call "belief" the entire operation authorized by figure 4. We understand once again that belief in no way refers to a cognitive ability. Instead, it refers to a complex configuration in which the Moderns construct themselves: in order to understand their own actions, they forbid themselves to return to fetishes even though, as we shall see, they continue to use them.

How Facts and Fetishes Blend Their Properties, Even among the Moderns

Far from explaining the fetishists' attitude, and far from justifying that of the anti-fetishists, belief allows two opposite and even contradictory repertoires of action to be kept separate. Together, these two categories try to account, clumsily, for what the Gold Coast Blacks calmly asserted: that they were building something that went beyond them. But the Moderns never use that distinction, not even to produce hard sciences, however much they appear to insist on it. As soon as the apparatus of belief is suspended, we notice that researchers all speak just like the Blacks whom the Portuguese silenced a little too quickly.

Let us listen to Louis Pasteur—one proof-worker among others—as he speaks, not about facts and fetishes, but about what is taking shape in his laboratory. If we apply the definition of belief given above, we should challenge him to choose between constructivism and realism. Either he has socially constructed his facts out of whole cloth (and thus he is adding no reality to the world's repertoire other than that of his fantasies, prejudices, habits, or memories), or else the facts are real (but then he did not construct them out of whole cloth in his laboratory). This contradiction seems so fundamental that it has been keeping the philosophy of science busy for three centuries without a break.

Yet it scarcely troubles Pasteur, who obstinately persists, as did the Blacks, in not understanding the challenge; he does not even perceive the difficulty. He asserts in one and the same breath that the ferment of his lactic acid is real because he carefully set the stage on which the ferment revealed itself on its own. The realists wax indignant: "You grant too much to the constructivists by admitting that you did it all yourself!" The social constructivists are similarly outraged: "How can you claim that lactic acid ferment exists on its own, without you, when you yourself are pulling all the strings!" And Pasteur persists, obstinately, gently, just like the old woman with a shaved head entering the Candomblé to seat, or make, or fabricate her divinity:

> All through this memoir, I have *reasoned* on the basis of the hypothesis that the new yeast is organized, that it is a living organism and that its chemical action on sugar corresponds to its development and organization. If someone were to tell me that in these conclusions I am *going beyond* that which the facts prove, I would answer that it is quite true, in the sense that *the stand I am taking is in a framework of ideas* that in rigorous terms *cannot be* irrefutably demonstrated. Here is *the way I see it*; whenever a chemist makes a study of these mysterious phenomena and has the good fortune to bring about an important development, he will *instinctively be inclined* to assign its primary cause to a type of reaction *consistent with* the general results of his own research. It is the *logical* course of the human mind in all controversial questions.[15]

It would be hard to be more constructivist. Thomas Kuhn or Harry Collins might have written such sentences, in which one can clearly perceive the scholar's effort to construct his facts by projecting onto them his professional habits, his presuppositions and even his prejudices, the habits of the group he belongs to, his bodily instincts, and the logic of the human mind. Unfortunately for the sociologists of science, Pasteur goes right on to add the following:

> And it is my opinion, at this point in the development of my knowledge of the subject, that *whoever* judges *impartially* the results of this work and that which I shall shortly publish *will recognize with*

me that fermentation appears to be correlative to life and to the organization of globules, and not to their death or putrefaction. Any contention that fermentation is a phenomenon due to contact in which the transformation of sugar takes place in the presence of the ferment without giving up anything to it or taking anything from it, *is contradicted by the experiment* as will be soon seen. (ibid., emphasis added)

Traitor! He has changed his philosophy of science all of a sudden. The constructivist has become a realist of the dullest and most ordinary sort. The facts speak for themselves in the eyes of impartial colleagues!

Has Pasteur contradicted himself? He has, from the standpoint of critical thought, but not from his own standpoint, and thus not in ours. For him, constructivism and realism are synonymous terms. Facts are fabricated, as we have known since Gaston Bachelard; but critical thought had trained us to see the fetishism of the object in this ambiguous etymology. Whereas we fabricate them in our laboratories with our colleagues, our instruments, and our hands, facts are supposed to become, by some magical effect of reversal, something that no one has ever fabricated before, something that holds up against any change in political opinion, any torment of the passions, something that stays put when someone pounds the table with his fist and shouts: "Here are the facts! They're not going away."[16] After the work of construction, the anti-fetishists claim, the facts "take on their own autonomy." Even though the very same French word, le fait, means both "what somebody has fabricated" (the manufactured thing) and "what nobody has fabricated" (the autonomous fact), must we see in this a contradiction that has first been covered up by a magic trick, then hidden by belief, before finally being buried under hypocrisy?[17] Not necessarily. Another solution is at hand, but this one presupposes that we can abandon critical thought, forget notions of belief, magic, hypocrisy, and autonomy, losing the stunning mastery that has made us Moderns and proud of it.[18]

The new repertoire appears as soon as we circumvent anti-fetishism and transform it so that it stops being the essential resource of our intel-

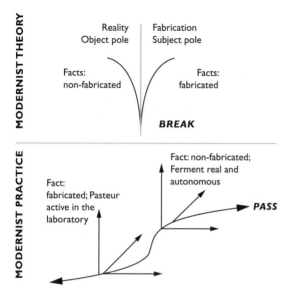

MODERNIST THEORY

Reality   Fabrication
Object pole   Subject pole

Facts: Facts:
non-fabricated   fabricated

BREAK

MODERNIST PRACTICE

Fact: non-fabricated;
Ferment real and
autonomous

Fact:
fabricated; Pasteur
active in the   **PASS**
laboratory

FIGURE 5.
In the modernist theory, scientists have to choose between fabrication and reality (first break). In the modernist practice, they find ways to *obtain* autonomy *because* of their active fabrication (no break). The two repertoires are kept strictly distinct (break between top and bottom).

lectual life and becomes instead an object of study for the anthropology of the Moderns. The first repertoire obliges us to choose between the two meanings of the word fact: is it constructed (a fabrication), or is it real (a fact)? The second repertoire accompanies Pasteur when he finds synonymy in the sentences "Yes, it is true, I made it in the laboratory," and "therefore only the autonomous ferment appears to the eyes of an impartial observer."[19] Whereas the Modern repertoire (top of fig. 5) keeps anything from happening in the middle (except for some tripping over the twists and turns of dialectics), in the non-Modern repertoire everything happens in the middle, such as that crucial moment when Pasteur, because he had done a good job, could let his ferment go, finally autonomous and visible, feeding delightedly on the culture that had just been invented for it. While the notion of fact is broken into two parts at the top of the diagram, at the bottom it serves as a means of passage to establish continuity between human work and the ferment's independence. The laboratory thus brings fact-making into play. The double articulation of the laboratory and Pasteur allows fact-making to "make talk," thus bringing us back to the two etymologies of the words "fetish" and "fact." The laboratory becomes, so to speak, the prosthesis allowing the ferment of lactic acid to speak as well as Pasteur; it allows the articulation between Pasteur and "his" ferment, between the ferment and "its" Pasteur.

The decisive importance of science studies, or the anthropology of sciences, is becoming clear. These disciplines act as a veritable clinamen, a slight deviation breaking the invisible symmetry that allowed belief to exercise its rights.[20] In fact, by forcing theory to take the practices of researchers into account, the social analysis of the sciences mixes the two repertoires and makes it necessary to explain the acknowledged facts of science with resources that were developed for accounting for fetishes![21] Of course it fails. One can hardly explain black holes by using the first critical denunciation, which was invented for use against fetishes and gods. But the very failure of these explanations has gradually unraveled all critical thought. When such explanations are applied to "real objects," the congenital weakness of the first denunciation quickly comes to light; but the process also brings to light, symmetrically, the powerlessness of objects—objects that have been disputed, socialized and tangled up in their (social?) conditions of production—to serve as hammer and anvil for the causal determination of human intentions. Social explanations may not have been worth much, but objective causality was no better. We had to go back to square one and start listening again to what the ordinary actor was saying.

Here is a *felix culpa* that allows us to stop believing in the essential, radical, foundational difference between facts and fetishes. What good is this difference, if it does not even allow us to account for scientific work?[22] Why insist so strongly on an absolute distinction that can never be applied? Because it is used precisely to redouble the advantages of practice with those of theory. The double repertoire of the Moderns cannot be discovered in their distinction between facts and fetishes, but in the *second*, more subtle distinction between the separation of facts from fetishes that they institute in theory, on the one hand, and the totally different passage from one to the other that they carry out in their practice, on the other. Belief then takes on a new meaning: it is what allows one to keep the practical form of life, in which one causes something to be fabricated, at a distance from the theoretical forms of life, in which one has to choose between facts and fetishes. This is how critical thinkers can go on purifying theory indefinitely, without having to face the possible consequences of their purification.

How the Practical Competence of Factishes Eludes Theory

As soon as we begin to size up practice, we notice that the ordinary actor, Modern or not, starts speaking in exactly the same terms as the coastal Blacks and the Candomblé initiates, in whose company I began this brief discussion. The ordinary actor blurts out something entirely self-evident, namely that his construction has gotten slightly out of hand. "We are indeed manipulated by forces that go beyond us," he might say, tired of being tossed about in all directions and accused of naiveté. "It hardly matters whether they are called divinities, genes, neurons, economies, societies, or affects. Maybe we were wrong about what word we should use for these powers, but not about the fact that they count more than we do. Conversely," the ordinary actor might continue, "we were indeed right when we said that we fabricate our own fetishes, since we are at the origin of the various forces that you want to take away from us, by making us into puppets, manipulated by the powers of the market, evolution, society, or the brain. Maybe we were wrong about what name to give our freedom, but not about the fact that we act in concert with others, whether we call them divinities or aliens. What we fabricate never possesses, and never loses, its autonomy."

The words "fetish" and "fact" have the same ambiguous etymology— as ambiguous for philosophers of science as for the Portuguese. But each of the two words emphasizes the inverse nuance of the other. The word "fact" seems to point to external reality, and the word "fetish" seems to designate the foolish beliefs of a subject. Within the depths of their Latin roots, both conceal the intense work of construction that allows for both the truth of facts and the truth of minds. It is this common truth that we must bring out, without believing either in the ravings of a psychological subject, steeped in daydreams, or in the external existence of cold and ahistorical objects that fall into a laboratory as if dropped down from

heaven. And without believing in naive belief either. Joining the two etymological sources, we shall use the label *factish* for the robust certainty
that allows practice to pass into action without the practitioner ever believing in the difference between construction and reality, immanence
and transcendence.[23]

As soon as we start to consider practice in this way, without worry-
ing any longer about having to choose too quickly between construction and truth, all human activities—not only those of Candomblé initiates or laboratory researchers—begin to speak about the same passage,
the same *factish*. Don't novelists, too, say that they are carried away by
their characters? To be sure, they are accused of bad faith. After an initial interrogation—"Do you fabricate your books? Are you fabricated by
them?"—they answer, obstinately, just as the coastal Blacks and Pasteur did, with one of those marvelous formulas whose meaning always
threatens to elude us: "We are the offspring of our works." And let no
one try to tell us that the practitioners are engaging in dialectics, and that
subjects—auto-positioning themselves within objects—are revealed to
themselves by alienating themselves via objects; artists, who could not
care less about subjects or objects, in fact move back and forth between
them, without at any moment even lightly brushing either the subject-
master-of-his-thoughts or the alienating object.[24] All of us who sit down
in front of a computer keyboard know that we find out what we think
about things by reading what we were writing. But we also know that
we cannot be submerged, for all that, in word play or in a Zeitgeist that
would speak to us unawares, for the excellent reason that such second-
class manipulators would have no more mastery over us than the author
has over his or her text. This everyday experience is made incomprehensible by the double suspicion of criticism, and that is why it is relegated
to the half-silence of simple practice.

Why should one require Gold Coast Blacks to choose between the
human fabrication of fetishes and their transcendent truth, while we
Whites, we Moderns, never choose, unless we are tortured and forced to
break up with our own hands the continuous passage that we have just
explored in practice?[25] In all our activities, what we fabricate goes be-

yond us. On the same basis as novelists, researchers, or sorcerers, so too politicians are challenged to lie on the Procrustean bed, unless they wish to be taken for liars.

> "Are you constructing a national representation?"
>
> "Yes, of course," they answer, "and from whole cloth."
>
> "So you are inventing, through manipulation, propaganda, and trickery, what the representatives should say?"
>
> "No, we are faithful to our constituents, because we are constructing the artificial voice that they would not have without us."
>
> "Blasphemy!" cry the critics. "We don't need to hear any more! Lost in their illusions, they can't even recognize their own lies!"[26]

And yet politicians, who have been reduced to silence for two centuries or more, pass back and forth every day, at all hours, between this artificial construction and this precise truth, just as researchers, who (at least according the textbooks) are expected to choose between construction and truth, spend every day and many nights in the laboratory constructing true truth.

The choice that the Moderns are proposing is thus not between realism and constructivism, but between that choice itself and a practical existence that grasps neither the terms of that choice nor its importance. Whereas before we could only swing violently back and forth between the two extremes of the Modern repertoire—or else attempt to go beyond these extremes using dialectics, as the Baron of Münchhausen "went beyond" the laws of gravity—we can now choose between two repertoires: the one in which we are challenged to choose between construction or truth, and the one in which construction and truth become synonyms. On the one hand we are paralyzed, like Buridan's ass having to choose between facts and fetishes, on the other we *pass* thanks to factishes.

In this way the ordinary actor, when questioned, will very explicitly and with wild intelligence multiply the life forms that allow one to pass, thanks to factishes, without ever obeying the either-or choice of the Modern repertoire. Yet these subtle theories will remain hidden, since the only way to sum them up officially lies in the choice that has to be made between construction and autonomy, subject and object, fact and fetish. Let us not try to simplify the situation: we cannot remain unaware either of the multiplicity of discourses that speak of passing while turning away from the Modern alternative, or of the importance of the Moderns' theory requiring a choice that never seems to serve any purpose. There is something sublime in the comparison between this immense surface—of discourses, mechanisms, practices, and subtle remarks, by which the actors-themselves assert the self-evidence of an easy passage on both sides of the words "fact" and "fetish"—and the fussy, Pharisaic attentiveness with which we believe this passage to be closed forever.[27] Or at least since we began trying to convince ourselves that we have been Modern, that is to say, radically and not just relatively different from the people of the Gold Coast.

Let us go even further. The very notion of practice arises from the requirement the Moderns impose. Since we cannot express ourselves in the either-or terms of critical thinking, we are obliged to continue doing what we have always done, but in secret.[28] Practice is the clandestine wisdom of the passage that stubbornly maintains (but since it can no longer say so out loud, it is content merely to act accordingly, to murmur its message in hushed tones) that construction and reality are synonyms. A strange clandestinity, since it is also, in our common experience, one of those secrets that everyone knows about, told in a thousand different ways through a thousand different channels. Yes, but theory continues— and for very good reasons that we must now grasp—not to take these multiple confessions seriously. We shall now use the word "belief" for the process that allows one to keep an official theory at the farthest possible distance from its informal practice, without any relationship between the two except for the passionate, anxious, and meticulous care taken to maintain the separation, and we shall call the anthropological description of that process "agnosticism."

How to Portray an Anti-Fetishist

In order to grasp the mysterious effectiveness of this separation between theory and practice, we would need to have portraits of anti-fetishists at hand; then we could counter-analyze the Moderns by preparing an ethnographic description of their iconoclastic gestures. Since to my knowledge no studies of this sort have been carried out to date,[29] I have chosen an enlightening anecdote from a contemporary Indian novelist.[30] Jagannath is a Brahman of the modernizing kind. He wants to break the fetishes and liberate the pariahs, who work for his aunt, from their alienation by forcing them to touch the holy black stone, his ancestors' shaligram. One late afternoon after work he seizes the stone from the altar, and in front of his horrified aunt and the priest, starts to carry it over to the serfs gathered in a corner. But in the middle of the courtyard Jagannath hesitates at the prospect of what he is about to do, stops, and questions himself:

> Words stuck in his throat. This stone is nothing, but I have set my heart on it and I am reaching it for you: touch it; touch the vulnerable point of my mind; this is the time of evening prayer; touch; the *nandadeepa* is burning still. Those standing behind me [Jagannath's aunt and the priest] are pulling me back by the many bonds of obligation. What are you waiting for? What have I brought? Perhaps it is like this: This has become a *shaligram* because I have offered it as stone. If you touch it, then it would be a stone for them. Thus my importunity becomes a *shaligram*. Because I have given it, because you have touched it, and because they have all witnessed this event, let this stone change into a *shaligram*, in this darkening nightfall. And let this *shaligram* change into a stone. (100–101)

But to the great surprise of Jagannath, breaker of idols, liberator, and anti-fetishist, the pariahs back away, terrified. He remains alone in the middle of the courtyard with an object that is now half-stone, half-divinity, the priest and his aunt crying out in shame behind him, while those whom he wanted to free huddle as far away as possible from the sacrilegious sacrificer.

Jagannath tried to soothe them. He said in his everyday tone of a teacher: "This is mere stone. Touch it and you will see. If you don't, you will remain foolish forever."

He did not know what had happened to them, but found the entire group recoiling suddenly. They winced under their wry faces, afraid to stand and afraid to run away. He had desired and languished for this auspicious moment—this moment of the pariahs touching the image of God. He spoke in a voice choking with great rage: "Yes, touch it!"

He advanced toward them. They shrank back. Some monstrous cruelty overtook the man in him. The pariahs looked like disgusting creatures crawling upon their bellies.

He bit his lower lip and said in a firm low voice: "Pilla, touch it! Yes, touch it!"

Pilla [the overseer] stood blinking. Jagannath felt spent and lost. Whatever he had been teaching them all these days had gone to waste. He rattled dreadfully: "Touch, touch, you TOUCH IT!" It was like the sound of some infuriated animal and it came tearing through him. He was sheer violence itself; he was conscious of nothing else. The pariahs found him more menacing than Bhutharaya [the demon spirit of the local god]. The air was rent with his screams: "Touch! Touch! Touch!" The strain was too much for the pariahs. Mechanically they came forward, just touched what Jagannath was holding out to them, and immediately withdrew.

Exhausted by violence and distress Jagannath pitched aside the *shaligram*. A heaving anguish had come to a grotesque end. Aunt could be human even when she treated the pariahs as untouchables. He had lost his humanity for a moment. The pariahs had seemed to be meaningless things to him. He hung his head. He did not know when the pariahs had gone. Darkness had fallen when he came to know that he was all by himself. Disgusted with his own person he began to walk about. He asked himself: when they touched it, we lost our humanity—they and I, didn't we? And we died. Where is the flaw of it all, in me or in society? There was no answer. After a long walk he came home, feeling dazed. (101–2)

The blow that Jagannath intended for the fetish—the idol, the past, the chains of servitude—has slipped. What now lies broken, dispersed, is not the fetish, but humanity, his own and that of the pariahs, his aunt's, the priest's. He thought he was breaking the fetish, but it was the factish that broke. At that moment he became a "savage beast," and the pariahs, "disgusting creatures." The stupid objectivity of the stone, what Jagannath wanted to have them feel with their own hands, passed into the serfs, who became "things stripped of significance." Inverting the magical gifts of Midas in order to desacralize the shaligram, Jagannath made it an object that turns all who touch it into stone. He wanted to dispel the illusion of the gods, and—bitter irony—here he stands, "more menacing than Bhutharaya." He only gets the pariahs to obey him, finally, because they give in to the coalition of menacing divinities, those of the master added to that of the spirit-demon. Even then, the serfs only obey him mechanically. Beasts, objects, machines: they go through all the shades of inhumanity. Even more seriously, master and serfs "died" because the factish, which once broken can no longer hold, from the outside, what made them human. "Where is the flaw?" Jagannath asks himself. Does the human no longer live in the subject liberated from his chains, in the breaker of idols, in the modernizer with his hammer, but elsewhere, slightly elsewhere? Must one truly stay in the shadow of factishes in order not to die, not to become beast, stone, animal, machine? Does one merely need a stone in order not to become hard and cold like a stone?

By choosing the wrong target, the modernizing Indian in us teaches us much about himself, but mostly about the Whites. This is the lesson we need to learn.[31] If they are to be scholars, creators, politicians, cooks, priests, initiates, operators, artisans, butchers and philosophers, the Moderns have to pass, as everyone must, from construction to autonomy. If they lived without factishes, the Whites could not live; they would be machines, things, savage beasts; they would be dead.

All the same, they are not required to believe in fetishes; they do not have to follow the horrible scenography of anti-fetishism, as if they were attributing souls to stones. Everyone agrees that the saligram is just that, a stone and nothing but; only the denouncer, the breaker of idols, does not know this. He finds out too late. He misinterprets the priest's horrified shouts, and his aunt's. Jagannath believes they are watching, aghast, a liberating blasphemy, but it is for him, and him alone, that they feel covered in shame. How could he attribute such awful feelings to them? How could he accuse them of stone worship, such a monstrous idolatry? The priest, the aunt, and the pariahs already know what Jagannath discovers by his failure: it has nothing to do with belief, it is all about behavior. It is not about a fetish-stone, but about factishes, about those off-center beings that allow us to live, that is, to pass continually from construction to autonomy without ever believing in either. Thanks to factishes, construction and truth remain synonymous. Once broken, they become antonyms. We can no longer pass. We can no longer create. We can no longer live. Then we have to set up factishes all over again.

Thanks to Jagannath, their efficacy now becomes clearer. We started with the either-or choice that required us to decide whether we constructed facts and fetishes or whether, on the contrary, facts and fetishes allowed us access to realities that no one has ever made. We have observed that in actual practice this choice is never made by anyone, and that everyone passes somewhere else, on the sly and effortlessly, attributing both autonomy and human origins to the same beings in the same breath. To use the language of philosophers, no one has ever been able to distinguish between immanence and transcendence. But this stubborn refusal to choose always shows up, we now understand, as a simple practice, as something that can never be spoken or theorized, even if the "actors-themselves" keep on saying it and describing it in luxurious detail.[32]

The misdirected blow of the idol-breaker, like the *felix culpa* of science studies, will allow us to deflect the course of anti-fetishism definitively in order to describe the underpinnings of belief from the outside. Symmetrical anthropology now has an operator, the factish, that will help it resume its work of comparing, but without losing its way in the meanders of cultural relativism and without believing any longer in belief. By pushing agnosticism to this point, we no longer need to oppose fetishless Moderns who unveil before the eyes of Blacks or pariahs either an undisguised external reality or the abyss of their own internal representations. We no longer need to mock the Moderns who might believe in anti-fetishism just as naively as the Gold Coast Blacks believed in their fetishes and the old aunts in their saligrams. The Moderns too have a factish: a fascinating, subtle, sly trickster. Now we need simply to sketch out its form and understand how it works.

How to Draw the Moderns' Split Factishes

People sometimes make fun of the crude nature of fetishes: rough-hewn tree trunks; clumsily carved stones; grimacing masks.[33] I trust I shall be forgiven, then, if I propose a similarly awkward portrait of Modern factishes, a rather crude Macintosh diagram (fig. 6). What is particularly interesting about our factishes is that they have been doubly *broken*, first vertically, then laterally. The first break allows for a violent separation between the subject and object poles, between the world of representations and the world of things. The second, slanted break introduces an even more violent separation between the theoretical form of life, which takes this first distinction between objects and subjects seriously, and a quite different practical form of life in which we carry on in peace and quiet, without ever being able to make a definitive distinction between what we make with our own hands and what exists outside of our hands.[34]

Faced with the cleverness of this mechanism, we can understand how the Moderns can believe that they and they alone are free from belief and fetishes. On top, the break between constructing subjects and autonomous objects hides the factish. On the bottom, the effectiveness of the factish is displayed, but the indefinite discourse referring to this effectiveness is constantly interrupted and displaced, as if it had to encrypt

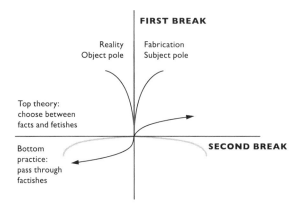

FIGURE 6.
The notion of *fac-tish* makes visible that the two breaks (the first between reality and fabrication, and the second between theory and practice) are constantly ignored by the middle pass.

the unending work of its mediations in order to make them invisible to theory. Between the two, the separation is total: this protects both the effectiveness of the passage on the bottom, and the purity of theory on top. The Moderns' factish is thus triply invisible, as long as others elsewhere, like Jagannath, do not reflect that unified image back to us. As soon as we grasp the image, the identifying picture, we notice that the factish resides in the mechanism as a whole. This mechanism must thus be portrayed in its entirety if we are to understand why the Moderns believe in belief and believe themselves to be fetish-less.

Wherever the Moderns must simultaneously construct and be caught up by what excites them—in public squares, laboratories, churches, courtrooms, supermarkets, asylums, artists' studios, factories, bedrooms—we have to imagine that such factishes are erected the way, earlier, a cross or the statue of an emperor might have been raised. But like the statues of Hermes castrated by Alcibiades, these factishes are all broken, shattered by the hammers of critical thought whose long history takes us back: to the Greeks who abandoned the idols of the Cave but put Ideas on pedestals; to the Jews who broke the Golden Calf but built the Temple; to the Christians who burned pagan statues but painted icons; to the Protestants who whitewashed frescos but brandished the true text of the Bible from the pulpit; to the revolutionaries who overthrew the old regimes but founded a cult devoted to the goddess Reason; to the hammer-wielding philosophers who put a stethoscope to the cavernous emptiness of the statues of every cult but put the ancient pagan gods of the will to power back in place.[35] As we can see in Mantegna's two "Saint Sebastians," in Vienna and Paris, the Moderns can replace the ancient idols that lie broken at their feet only with another statue, it too made of stone and on a pedestal, it too broken by the martyr, pierced by arrows, destroyed at once. Chase the factish through the door; it will come back through the window.

But no, I am mistaken. We have to add one more thing to these fac-tishes. We have to revise the diagram in order to incorporate the work by which the broken statues have been repaired, mended, patched up. We know that ethnologists and ethnopsychiatrists alike admire, and rightly so, the nails, hair, feathers, cowry shells, scarifications and tattoos that marked the old fetishes—I mean the destitute fetishes of the Gold Coast peoples before they were tossed into a bonfire or kept in a cabinet of curiosities. What can we say, then, about the extraordinary proliferation of marks, bits of string, feathers, barbed wire, scotch tape, pins, and staples that have always been used to repair the split top of the modern factishes, as well as the fasteners that used to keep them upright on their pedestals? From the beginning, everyone has done repair work, endlessly mending the two-fold tear.

Why do ethnologists take so little interest in this marvelous mend-ing that allows the effectiveness of factishes to be repaired every day in a thousand different ways, even though theory has broken the passage between construction and reality? If the factishes had been broken for good, no one, anywhere, could continue to act. But if the factishes had not been broken by solid hammer-blows, the Moderns would not be radi-cally different from anyone else. There would not even be any difference between the bottom of their factishes and the top. The Moderns would go into action as they have always done in darkest Africa, as they do every day in the vast chattering and silent country of practice. Why that bizarre configuration? Why break and then repair, as the Korean whose text I in-vented in the Prologue was wondering? Because, by surreptitiously carry-ing out the task of resolving the continuous contradiction imposed by the violent break of the passing and mediating factishes, the Moderns have been able to mobilize incredible forces that never looked menacing or monstrous. The broken top of the fetishes is not just one more illusio, an ideology that would disguise the true world of practice with a false consciousness. It upsets the theory of action, creates the independent world of practice, and allows it to unfold without ever having to justify itself on the spot. Thanks to the broken idols, it is possible to innovate without taking any risks, without accepting any responsibility or being exposed to any danger. Others—later, elsewhere—will pay the price, will feel the impact, evaluate the fallout, and try to contain the damage.

The researcher at the Institut Pasteur who introduces himself to me innocently, saying "Hello, I'm the coordinator for beer yeast chromosome 11," is doing exactly what astonished Lévi-Strauss when he heard a Bororo declare that "Bororos are Araras [that is, parrots]." The researcher is mixing up his own properties with those of beer yeast, just as Pasteur mixed up his own body with acid, and as the peoples of the Amazon region mixed up their own culture with the nature that surrounded them.[36] Of course, our researcher does not really mistake himself for a chromosome, any more than the Bororos really think they are parrots. But after the interview, after spending three hours going over Europe, the beer industry, Macintosh programs for visualizing DNA bases, and the *Saccharomyces cerevisiae* genome, he confides just as innocently: "But all I'm doing is science!" Here is the little difference, the break in symmetry. For if the world of parrots cannot shift without upsetting that of the Bororos and vice versa, it is possible for this researcher to take himself for a chromosome and upset a whole industry, a whole science, as if this double shock stirred up only homogenous facts. When beer yeast chromosome 11 appears in the world, it will simply add one element to the furnishings—all at once, by surprise—of nature alone, on top, in plain sight. On the other side, caught by surprise, others will suddenly have to take care of the consequences—ethical, political, and economic—of this action. The researcher himself does, will have done, will do, "only science."

Deep in your laboratory you can revolutionize the world, modify genes, reformat birth and death, implant prostheses, redefine the laws of the economy: all this will never appear as anything other than simple, silent, opaque practice. On top, in the light of broken fetishes, people will speak only of science on one side and of freedom on the other, without ever mixing the two—even if by some prudent mending and with many feedback loops, arrows, and round trips, the two broken parts are fitted back together, though their spirit can never be glued back into place: all the advantages of criticism on top; all the advantages of practice on the bottom; all the advantages of the meticulous distinction between the two; all the advantages of the passage from one to the other, with full (practical) knowledge of the three repertoires: the break, the passage, and the repair.[37] You can surely see that the Whites, too, are worthy of interest, since they are able to display enough distinctive features for comparative anthropology to study.

Let there be no misunderstanding here. I am not reducing the Moderns—in order to shame them for a moment—to the monstrous and barbaric piety with which they thought they had definitively broken. I am not reviving the theme of the idols of the forum, marketplace, or temple in order to accuse reasonable people of believing, in spite of everything, the way the Gold Coast Blacks or the pariahs did. I am not encouraging them to make an ultimate, heroic effort, like the hammer-wielding philosophers, and smash the last superstitions that might still slumber within the sciences and democracy. The very definitions of monster, barbarism, idols, hammer, and rupture have to be taken up again from scratch. There have never been any Barbarians; we have never been Modern, even in our dreams—especially in our dreams! If I am putting amulet-covered Portuguese on the same level as amulet-covered Guineans, fetishists on the same level as anti-fetishists, saligram worshippers on the same level as iconoclastic Brahmans, I am not dragging them all down, *I am pulling them all up*. Who knows most about paying equal respect to facts and fetishes? Why, obviously, those who have always gotten their factishes to say that they are being used as passages for what goes beyond them when they are being constructed. Are we Moderns, too, capable of such dignity? Of course we are, not to worry: otherwise you couldn't pray, create, think, discover, build, fabricate, work, or love. It is only that our distinctiveness comes from one particular characteristic feature: our factishes, even though they are broken, are mended in such a way that they relegate to practice what theory can grasp only in the double form of breakage and repair. Such is our tradition, that of fetish-breakers and menders of fetishes; such are our ancestors, to be respected, but not too much, as is the case for any venerable tradition.[38]

I must confess that I rather like the portrait of the Modern world that we produce if we reestablish the multitude of broken, mended, practical factishes at all points, in all places, on every steeple, gable, and temple, at every branching and intersection. We no longer need to contrast the disenchanted, virtual, absent, deterritorialized world with the other one: the rich, intimate, compact and complete world that belongs to the Primitives, who have never lived in the fetal quiet of noble savage dreams. But we no longer have to imagine, either, that we are going to get away—with our truth, efficiency, and bottom lines—from the horrible barbaric magma toward which, if we are not careful, our past might drag us down. There are no Barbarians, no savages; we Moderns, with our sciences and our technologies, our rights, our markets, and our democracies, are not barbaric either, whatever the Heideggerians may imagine.[39] We are like everyone else (where is the problem? where is the loss? where is the danger?), except that we are linked by countless ties to the particular factishes—our ancestors, our traditions, our lineages—that allow us to live and pass. We are indeed the heirs of those fetish-breakers and fetish-menders. Comparative anthropology now has the means to rekindle a dialogue that seems more fertile to me than the ones we hear at UNESCO, or the ones offered by the tedious resentments of anti-imperialism. For the first time, perhaps, we no longer have any barbarians—not outside the gates, and certainly not inside. For the first time, perhaps, when we use the word "civilization," this admirable term need not be surrounded by dark forces that are only waiting for the signal to rush across the dividing line and destroy everything. For the first time, perhaps, we can remember that civilizations are not mortal.[40]

Part Two
Trans-fears

We can now define anti-fetishism with precision: it is the prohibition on understanding how one passes from a human action that fabricates, to the autonomous entities that are welcomed by that action and revealed through it. Conversely, we can define symmetrical anthropology as that which lifts the prohibition and gives the factish a positive meaning. The factish can therefore be defined as the *wisdom of the passage*; as that which allows one to pass from fabrication to reality; as that which gives *an autonomy we do not possess to beings that do not possess it either*, but that by this very token give it to us. The factish is a fact-maker, a talk-maker. "Thanks to factishes," as sorcerers, initiates, researchers, artists, and politicians might say, "we can produce slightly autonomous beings that somewhat surpass us: divinities, facts, works, representations." Unfortunately, this way of speaking recycles familiar concepts such as "us," "production," "autonomy," and "surpassing," which have been forged over the centuries in order to nourish the very anti-fetishist polemic whence we are trying to extirpate ourselves.[41] After investigating the avatars of the object at some length, and after verifying that the object never occupies either the fairy-object position or the cause-object position, we must now turn to the avatars of the subject. Social constructivism obliges us to have as many misconceptions about its presumed tireless laborer as it does about the entities it mobilizes. If Pasteur can say, without contradicting himself, that he has turned lactic acid ferment into an autonomous entity, if the Candomblé initiate can unhesitatingly claim that she has to learn how to fabricate her divinity, if Jagannath's aunt can say without batting an eye that the saligram is nothing but a stone and that for that very reason the stone can give them all life, then the subject understood as the source of action must change as much as the target-object. I needed a site other than a laboratory to work out this notion of subjects, which is the counterpart of the notion of facts. Tobie Nathan offered me such a place, and I am going to try to do it justice — but I shall surely fall short.

How to Obtain Black Market Divinities
with the Help of Suburban Migrants

The setting is in the suburbs, in a sort of circle formed by psychiatrists, psychologists, students, ethnologists, visitors, journalists, onlookers, hecklers, and passers-by who participate in the session. Within this ring sits the patient, one link among others, neither privileged nor inferior. He is called a "patient" so he can fill out the insurance papers, but he does not really deserve the label, as he is very active. At all events, there is a vast difference between this situation and the patient rounds in the mental institutions I was familiar with, when philosophers still had to get a degree in psychology. The patient is here, to be sure, and his illness has a tight hold on him, but it will quickly loosen its grip and no longer deserve to be called an "illness," exactly. The patient—since we are stuck with this label—comes with his closest relatives (uncle, mother, father, brother, children), but also with his extended support group: judges; social workers; psychologists; educators. The former are often Black or dark-skinned; the latter are almost always White.

The patient speaks his own language(s). A first translator comments in French, then everyone has a go at translating. There is some surprise that the patient should not be at the center of the room, or of the conversation. Some try, indeed, to speak of him, to endow him with an inner life, a personal history, responsibility: "He's doing better; he's gotten himself under control; he's open; he's communicating." This seems to interest the others very little, however. They are looking—underneath, above, to the sides, elsewhere—and they speak of quite different matters: of what? Of divinities. At first, the patient is surprised, uncomfortable. Worn down by dozens of psychological (or perhaps we ought to say psychogenic) interviews, he seems bothered to be talking about that: about the id? No, not at all, not even close. No one in this circle is trying to pass from the dining room into the kitchen, scullery, and cellar of psychology. Though we may have come to speak of the child, we devote two hours to the mother and the grandparents. Having come to heal the sister, we remain focused on the uncle who stayed behind in the old country. Trying to understand a crime committed by the son of an immigrant, we spend the morning exploring Allah's relationship with his father and grandfather.

The patient's discomfort does not last. After a while, interested, he opens an eye and begins to participate in the conversation as if it were about someone else—and it is an Other, indeed, who is the subject of this multilingual conversation. Sometimes he contributes a comment. Sometimes even (to my astonishment as a moralizing and psychologically conditioned observer), everyone—including the patient—bursts out laughing over the terrible dramas that are being plotted around him. Are we all in a mental institution, ready for straitjackets when we leave? No, because in Saint-Denis, in France, we are observing a curious experiment: what psychological interviews can do, an ethnopsychiatric session can undo. The responsible and ill subject—we have known this since Foucault—has not existed from the beginning of time. In order to grab hold of him, to pin him down, careful formatting, vast and solid institutions, exercises of punishment and inquisition are required. But if these experimental conditions are modified, if the "patient to be psychologized" is thrown into a session at the Georges Devereux Center, then suddenly he becomes transformed into an entirely different agency. It is as if, over the course of three hours, we were witnessing the progressive liquefaction of the psychological subject, as it gradually detached itself from the patient, migrated slowly to the center of the session and finally dissolved there, only to reconfigure itself into a completely different shape. And the illness, finding nothing more to cling to, makes itself scarce as well, but no one really pays it much attention. As Freud used to say, the patient might be cured, but that will be a bonus.

Others can describe the sessions much better than I.[42] Since the treatment plan forbids impartial observation, I am going to talk about myself: an ignorant onlooker; patient and impatient; ill and healthy; compact and multiple. Not to worry, I am not going to put my own psychological makeup on display. No, I am going to take advantage of the experience to get rid of that too, as the session progresses; I am going to join in the gradual migration of souls, the process of detachment, in order to understand what White subjects are made of. How can one de-psychologize, in three hours, a patient burdened by forty-eight years of solid psycho-geneses?

But I should not be surprised. During three short hours in a labora-tory, some twenty years ago, I came to understand that all the objects of the exact sciences had to be "de-epistemologized." You must admit that the symmetry is awesome. At the Georges Devereux Center, migrants rediscover their divinities by losing their psychology; at the *Centre de Soci-ologie de l'Innovation* on the Boulevard Saint-Michel, researchers rediscover their collectives by losing their epistemology. I could not fail to make the connection. Two separate centers—unconnected except for the silent shuttle between them of a young woman, Emilie Hermant, and the wis-dom of a Belgian philosopher, Isabelle Stengers—were performing the same work, twice over: in one place on objects, in the other on subjects. What would Paris look like if I were to link the two centers, and if, to ob-jects that had been re-socialized by the new history of sciences, we were to add subjects to whom ethnopsychiatry had restored their divinities? There would no longer be any rational, efficient, profitable researchers trying to integrate migrants on the path to modernization into the Re-public. The multiple objects of the former would no more stay in place than the ancestors of the latter.

A patient (I, you, he) who, a moment earlier in the waiting room, was preparing for the examination of his deep or superficial self, finds him-self bound up by divinities of whose existence he was almost unaware, finding himself at the same time freed from the obligation of possessing an ego endowed with inner life and consciousness. As a participating observer, he watches others—who only pay him passing attention—as they interrogate aliens who are interested in him only by a fluke. Indeed, this is no longer about him. Perhaps he will be cured by it. But in order to understand this slippage, this sobering up, one would have to offer fetishes a dwelling place again; one would have to build a dovecote where divinities, like a flock of wood pigeons, could come back to roost, and coo to their heart's content. As the session progressed, I understood very quickly that it was not a matter of accepting the actors' "cultural representations," of entering into them with the condescending hypocrisy of psychologists and believing in the divinities on the pretext that the migrants believe in them (like the mentally disturbed patients in comic strips whose warders, in order to cure them, pretend to be Napoleon). Indeed, it is not a matter of believing or suspending one's ordinary beliefs. The divinities alone are acting: but how, and in which world, and in what form? Perhaps we are finally going to reap the fruits of our factishes. By modifying the definition of belief so profoundly, by pushing agnosticism so far, might I be able to situate this traffic in divinity more easily?

How to Do Without Interiority and Exteriority

It must be possible to make a place for the divinities again, provided that we modify the space in which they might be deployed. To do this, we have to redefine the shapes and voids that the notion of belief has drawn. Critical thought worked, if you will, as a gigantic suction and force pump. On the pretext that we help to fabricate the beings in which we believe, it emptied out all the fairy-objects, expelling them from the real world in order to transform them one by one into fantasies, images, or ideas. Critical thought drained space, creating emptiness. Conversely, on the pretext that fact-objects appear to exist without us once they have been worked over in the laboratory, critical thought lined up facts in tight ranks, forming a continuous, seamless "real world" without holes, without humans. Critical thought filled up space, creating fullness. By twice avoiding the curious practice that requires both fairy-objects and fact-objects to be fabricated by humans, this suction and force pump, operating by subtraction and by addition, by suction and by pressure, by emptying and by filling, simultaneously created both interiority and exteriority. There was no more room for divinities, but there were subjects galore, mistakenly thrown into a world of things. There was no more room for lactic acid, but there were external objects suddenly discovered by knowing subjects.

The work of belief creating interiority and exteriority allows us to understand better why we can no longer use psychology to situate subjects, any more than we can use epistemology to describe the peculiar history of objects. Neither can function without the other. Just as today's objects look nothing like the way we used to think they looked, when we believed that we knew—without intermediaries or mediations—today's subjects look nothing like the way we used to think they looked, when we believed we knew that naive belief existed somewhere. Belief and knowledge sailed, and sank, in the same boat. As the world was filled to bursting with known or knowable objective causes, and as certain primitive, archaic, infantile, or unconscious human beings obstinately tried to populate it with nonexistent fetish beings, those empty-headed fantasies had to go somewhere. Where to stuff them? Why, into the empty head, of course. It's full? No matter, let's hollow it out! Let's invent the notion of an interiority filled with hollow dreams, having no reference whatsoever to the reality known only to the exact or social sciences.

It is clear that the subject endowed with interiority serves as counterweight to objects endowed with exteriority. In order to link them, we have to invent the notion of representation. Thanks to this notion, the subject endowed with interiority begins to project its own codes onto "external reality," codes that supposedly come to it, via a most impressive causal chain, from linguistic structures, from the subconscious, from the brain, from history, from society. This time the confusion is complete. A subject that is the source of action, endowed with interiority and consciousness, arbitrarily carves out an external reality that exists independently from it and determines, through some other channel, those same representations. Such people dared to interrogate the Guineans? This is obviously a case of the pot calling the kettle black. Worse, determined not to repeat the condescension of the conquering Portuguese, a few pious hypocrites claimed that they were respecting the savages by raving just as the others did, and that the poor Blacks or Brahmans had the good fortune to possess their own "social representations" that divided up reality along different lines and according to other arbitrary impulses. Turning others into moved and grateful partners of Modernist ravings is a peculiar way of showing them respect! Cultural relativism adds one last delirium to all that have come before.

One could, of course, do without interiority altogether by naturalizing internal life. Indeed, critical thought offers a rich repertoire—too rich, too easy, too cheap—for drowning the subject in the objective causes that presumably manipulate him (see fig. 3, p. 14). There is nothing easier than making the subject a superficial effect of language play: the temporary capacitance emerging from a neural network; a phenotype of a genotype; the conscious of the subconscious; the "cultural idiot" of a social structure; a consumer of world markets. How to chop off the subject's arms and legs: we have all learned how to perform such amputations by reading the newspaper. We have been preparing ourselves for the death of mankind since we took a class in Theory 101. Fortunately, these procedures have been off limits to us ever since the little scandal mentioned earlier: the *felix culpa* of the anthropology of sciences. One would have to believe—either wholly or in part—in one or another of the many imported social or natural sciences, in all seriousness, in order to shut up the overly chatty speaking subject. But to pass brutally from autonomous subjects to the scientific objects that determine them would prolong anti-fetishism instead of leaving it behind. We do not want to drown Pasteur, with his attentiveness to the precise gestures that revealed his ferment, any more than we wish to lose our Candomblé initiate who is fabricating his own divinity, or fail to see how Jagannath's ancestors went about making what keeps them alive from a simple stone. Our theory of action must pick up just what Pasteur and the ancestors are doing that is uniquely theirs, at the very moment when their actions go slightly beyond them, precisely because their actions have gone slightly beyond them.

Curiously, the path of the factishes (bottom of fig. 6, p. 30) seems much simpler, more economical, more reasonable—yes, reasonable. Instead of giving ourselves (1) cause-objects filling every possible gap in the external world; (2) source-subjects endowed with interiority and crammed with fantasies and affects; (3) more or less arbitrary representations groping about, more or less successfully, in an effort to establish a fragile tie between the illusions of the ego and harsh reality as known by the sciences alone; (4) new causal determinations to explain the arbitrary origins of these representations; why should we not abandon the double notion of knowledge/belief, and populate the world with the tangled entities that emerge from the mouths of the "actors-themselves"?

Pasteur does not require his lactic acid ferment to be external to himself, since it is available to him in the laboratory, and since he even gave it—as he ingenuously admits—a bit of a boost, in terms of his own prejudices, to make it become a living being. And yet Pasteur does ask us to recognize that the ferment has all the autonomy of which it is capable. The Candomblé initiates in no way claim that their divinity speaks to them directly through some heavenly voice, since they admit, just as ingenuously, that their divinity is at risk—for want of "know-how" on their part—of becoming an "endangered species." And yet, coming from them, this confession does not weaken, but strengthens the very existence of the divinity that speaks to them. Jagannath's aunt does not ask the stone to be anything but a stone. No one, in practice, has ever displayed naive belief in any being whatsoever.[43] If there is such a thing as belief at all, it is the most complex, sophisticated, critical, subtle, reflective activity there is.[44] But this subtlety can never unfold if one first attempts to break it down into cause-objects, source-subjects, and representations. To take away the ontology of belief, on the pretext that it occurs inside the subject, is to misunderstand objects and human actors alike. It is to miss the wisdom of the factishes.

The Chinese have a proverb that applies wonderfully to the denunciatory attitude of critical thought: "When a wise man points to the moon, the fool looks at the finger." Instead of looking at what is attracting the actors' full attention, the anti-fetishist thinks he is very clever for denouncing, with a shrug, the object of belief. It does not exist, as he knows by omniscience, or rather by intuition; he directs his attention to the finger, then the wrist, the elbow, the spinal chord, and from there toward the brain, then the mind, from which he comes back down following the objective causalities offered by other sciences, toward education, society, genes, evolution: in short, the full world that subjects' fantasies cannot even begin to threaten. A much simpler, more intelligent, more economical, and finally more scientific hypothesis would consist in directing one's gaze, as the proverb says, not only toward the moon,[45] but also toward lactic acid ferments, divinities, black holes, tangled genes, apparitions of the Virgin, and so on. What do we have to lose? What are we afraid of? That the world might be too populated? It will never be populated enough. The emptiness of infinite space is probably what frightens us. Just as the scholastic world abhorred a vacuum, the world of social and causal explanations abhors the variable-geometry ontologies that might force it to redefine not only action but also actors, and that might unfold into deep space like planets and galaxies, each irreducible to the other.

Fortunately, the fear of insufficiently restricting the population of beings, the fear of abandoning the difference between epistemology and ontology, belief and knowledge, stems simply from the confusion produced by critical thought. The piston noise of the suction and force pump, and that alone, is what keeps us from noticing that the "actors-themselves" only rarely require beings with whom to share their life that they exist in the form of brute facts: continuous; persistent; and stubborn. When Elizabeth Claverie makes a pilgrimage to Medjugorje to see the Holy Virgin appear on the stroke of noon, she is not behaving like the fool in the Chinese proverb; she does not start out priding herself on her knowledgeable superiority, saying, "Since I know perfectly well that the Virgin doesn't exist, and doesn't appear, I'm only going to try to understand how some lowly French pilgrims could believe in its existence, and why."[46] She follows the finger pointing to the Virgin, a very wise and above all a quite empirical position. Yes, of course the Virgin appears, everybody sees her; the whole crowd crackles with Polaroid as everyone gets an imprint. Elizabeth sees her too: how could she do otherwise? But if we now listen to the multiform speech arising from the prayerful crowd, as well as to the emotional whispers on the train taking the pilgrims back to Paris, we notice that at no moment did the photographers expect to see the Virgin inscribed, like a Saint Sulpice statue, onto the silver bromide. In no way does the Virgin demand to occupy the position of thing to be seen or illusion to be denounced; Pasteur's ferment never really required at any point, in order truly to exist, the role of constructed object, or found object; the saligram never asks to be anything but a simple stone. The ontological envelope drawn by the Virgin who saves, her "list of specifications," as it were, obeys constraints that never cross between the two poles of impoverished existence and impoverished representation.[47] She is doing something else altogether: she is occupying the world in a way that surprises not only the priests but also the priest-haters.

Does the only example of naive belief we have, then, come from a naive belief on the part of researchers that ignorant people believe naively? Not quite, for ignorant people do exist who quite resemble the picture that researchers would like to paint of them. Photographers of flying saucers, archaeologists of cities lost in space, zoologists tracking the Yeti, people who have been contacted by little green men, creationists fighting against Darwin—all the sorts of people that Pierre Lagrange studies with a collector's passionate interest—are all trying to pin down entities that seemingly display the same properties of existence, the same specifications, as entities that, according to the epistemologists, come from laboratories.[48] Curiously enough, these people are called "irrationalists," whereas their greatest fault comes more from the reckless trust they display in a scientific methodology, dating back to the nineteenth century, in order to explore the only mode of existence they are able to imagine: that of the thing, already there, present, stubborn, waiting to be pinned down, known. No one is more positivistic than creationists or ufologists, since they cannot even imagine other ways of being and speaking than describing "matters of fact." No researcher is that naive, at least not in the laboratory. This is so much the case that, paradoxically, the only example of naive belief we have seems to come from the irrationalists, who are always claiming that they have overthrown official science with stubborn facts that some conspiracy had hidden away.

However, if we look more closely, we find that even that kind of scientism might escape the accusation of naiveté, for the endless quest of the ufologists aims at tangled objects—certainly impoverished ones—that do not manage to play the role that scientism had prepared for them. This is a strange misunderstanding, which would then leave naive belief without any examples to use as evidence. The result would be amusing. Epistemologists would thus display, to our eyes, the only definite case of naive belief in the first degree. A new cogito, a new fixed point: I believe in belief, therefore I am modern! Even that hapax is unproven, however, since the political will that keeps belief in belief alive despite the universality of all the counter-examples—thus reversing the principle of induction—portrays an interesting tangled object itself! There are many good political reasons to believe in the difference between reason and politics.[49]

How to Establish a List of Specifications for Divinities

From now on, when a finger points toward the moon, we shall look at the moon. Thought counts for less than thought-of beings; it is to these that we must become attached. Equipped with this outcome, let us try to return to the therapy session. At the time, I had no room to install the divinities without turning them into mere representations. But how can we claim to respect entities that we have first deprived of their existence? Is existence not among the perfections indispensable for respect, which the idea of belief never allows us to preserve?[50] Thus I had to come back to the crack that runs between epistemological questions and ontological questions. The new history of the sciences has allowed me to slip in between the two. The lactic acid ferment that Pasteur discovered, constructed, induced, and formed has served as my model for an understanding of the divinities. That ferment would not have its place in the world, either, if things had to be divided up into causes, interiorities, and representations. This is the advantage of symmetry: by taking the most respected beings of a culture—our own—as examples, we can shed light on the most despised beings of another culture. All ask to exist; none is caught in the choice—viewed as a matter of good sense—between construction and reality, but each requires particular forms of existence whose list of specifications must be carefully drawn up.

I have already fulfilled the first condition of these specifications: the divinities installed in the therapeutic context exist for real. Of course, I run the risk of immediately weakening that realization if I attribute exis-tence too generously. Indeed, at first glance we have too many things to take into account, since dreams, unicorns, and mountains of gold must be able to cohabit, without standing out in any way from gods, spirits, lactic acid ferments, works of art, societies, saligrams, genes, and ap-paritions of the Holy Virgin. Since we have voluntarily given up the re-sources offered by anti-fetishism, and since we can no longer categorize all these entities in the four lists of the critical repertoire (see fig. 4, p. 15), we get the heady feeling that "anything goes." Next to this onto-logical relativism, cultural relativism seems almost innocent. Like the Hebrews in the desert waxing nostalgic for the onions their Egyptian masters used to dole out to them, are we not going to regret the solid difference between psychic mechanisms, representations, and causes? At least it had the advantage of organizing all the clutter, forcing us to distinguish each time between what lay dormant in the interiority of sub-jects and what was lying around in the exteriority of objects. This new, overly slack ecumenicalism plunges us into a darkness in which all cows are gray. Horrified by the confusion, are we not tempted to turn back and raise the same questions once again, in the shadow of the factishes split off from the Moderns: is it built by us? Is it autonomous? Is it in our heads? Is it in things? Are we the masters, or are we being surpassed?

Before we regress to that stage, let us understand the advantage of this sort of ecumenism for our understanding of the therapeutic process. We do not give in to our irrational side when we accompany a patient who is mobilizing his divinities, any more than we give in to our rational side when we follow Pasteur getting mixed up with his ferment. There are no more sides—or at least there are not just two sides any longer—but many sides, which form facets, or folds. Wondering how these entities hold together, once they have been torn from the two solid hooks of sub-ject and object, is the same as wondering where the suns, galaxies, and the planets fall when we lose the Aristotelian cosmos. They hold up by themselves, provided that the referential framework of a finite world—endowed with a top and a bottom—no longer forces them to make some relative movement, either falling or climbing. Similarly, irreducible entities hold each other up quite well. They lie in their own world without excess or residue. If we grant this point, it becomes possible to speak fervently, warmly, and enthusiastically about Pasteur stirring up his laboratory, his career, and his ferment, and to contemplate coldly, distantly, and precisely the Candomblé initiates preparing their own divinities. Nothing stops us from using language games the wrong way round, since they no longer correspond to ontological domains: some of which are cold and others hot; some open and others closed; some spiritual and others materialistic.

To put it differently, the lower half of factishes does not introduce us to the fog of mystery. It is darkened only by the overhanging shadow of its top, which is the only part that can aspire to the light. Let us set this brightness aside! Our eyes rapidly get used to the phosphorescent light that seems to emanate from the entities themselves, as it does from the active matrices of flat computer screens that nothing illuminates from the outside. The language of mystery, the trembling voices, the quavering, the flusters—all that came from the awful transcendence that we had sought to add to the simple world known by science alone. Indeed, no longer able to place the countless entities with which we mix our lives (since the traditional image of science had depicted this lowly world as filled by effective causalities), and no longer able to resign ourselves to housing them in the folds of our egos to make them into fantasies, we had no other choice but to invent another world filled with gods, devils, poltergeists or succubi: exotic clutter; an asylum from gnosis; an attic filled with all the New Age baubles. To speak of mystery—or worse, to whisper mysteriously—would be to blaspheme against all the factishes: those of the divinities, of course, but also those of the laboratories. Should we divide the world into top and bottom, nature and supernature, we would prevent ourselves from understanding: both Pasteur and his ferment; the patient and his divinities; the pilgrim and his Virgin; Jagannath and his stone. There is no other world any more than there is an underworld. There is no reason to fall into the ego's fantasies, either. Once those three drains have been plugged, there is no longer any particular mystery; or at least the mystery becomes, like common sense, the most widely shared thing in the world. We are all, as it were, slightly overtaken by events.

How to Transfer the Fears

Once the divinities have been seated in existence, let us add to the specifications list the requirement that we be able to refer to them in precise and exact language without using any of the scenographies of exoticism, and without needing to believe that they have come from another world, different from our own: a world that is presumed, in contrast, to be flat, lowly, full, causal, and reasonable. What else must we add in order to distinguish the divinities' way of existing from that of other entities? Here I run into a new difficulty. Must I believe what the ethnopsychiatrists say about what they do, or should I follow their practices? Let us not forget that, with the Moderns, the upper and lower parts of factishes are totally opposed. What is true for philosophers of science may also be true for ethnopsychiatrists, yet the session I am participating in is held in the suburbs; it mixes diverse ethnicities that never would have met except on French soil; it takes place in several different languages; it is videotaped; it is reimbursed by the national health insurance plan; some of the migrants who show up have been integrated into their new homeland for quite a while; finally, there are also patients who are native-born French from the provinces. It would be hard to imagine a more heterogeneous, less traditional, setup. This interests me precisely because it is an artificial instrument. The fact that this apparatus—similar to a centrifuge or a calculator—is clothed in a whole folklore that speaks of cultures, authenticity, a return to the ancestors, village gatherings, baobabs, or traditional healers, is thus of little concern here. In what follows, I am interested in separating the effect produced by this long-range experimental instrument from the ethnography by which some seek to define it.[51]

One kind of energy in particular is produced, put together, spread out, worked, constructed, and dispatched in this context. How can this energy be harnessed? How can we define it? After removing any pretension to authenticity (which would be contradicted by the very nature of

innovation, and which would not allow a full appreciation of its originality), I must rule out another phenomenon, certainly an important
one, but one that blurs the interest of the energy that concerns us (I am
still speaking as a patient and an ignorant person). Practitioners strive,
through therapy, to give patients back an identity, to re-affiliate them, to
re-territorialize them. The fabrication of an identity, however, requires
vehicles, means, procedures, and arrangements other than the ones that
mobilize divinities. We Whites descended from monkeys are no less af-
filiated than those who descend from heroes, totems, or clans.[52] Soccer,
rock and roll, drugs, elections, salaries, and schools affiliate perhaps as
deeply as ancestors, race, the land, and the dead. At a minimum, in any
case, the construction and transformation of cultures are phenomena
that are too complex to be reduced to the substance of a definite identity
that one might find when one goes back home. Culturalism crumbled
long ago, along with the exoticism that conveyed it.[53] One can no longer
account for treatment by reviving this ghost. Its roots grow in too many
directions, reconnect too quickly, and form rhizomes whose branchings
are too surprising for us to hope to affiliate patients by treating them
as Baoulés, Kabyles, or Beaucerons by nature. Migration and the neo-
formation of new cultures, at this very moment all over the world, would
make such an attempt impossible in any case. Moreover, to imagine that
Gold Coast Blacks alone have such powerful cultures and deep ancestral
roots, while Whites wander soulless and free of the dead, would amount
to inverting Charles de Brosses's racism and a failure to apply the prin-
ciples of symmetry, which would be, for me, a cardinal sin.

Thus the theoretician of ethnopsychiatry is of less interest than the
practitioner. What does the latter do? He heals by means of gestures per-
formed within an experimental and artificial framework that reveals a
particular type of energy whose existence we had forgotten, preoccupied
as we were with epistemologizing our objects and psychologizing our subjects. He is a great "charlatan," and I would not have understood what he
was doing if I had not given a positive twist to this word, which is usually
used to stigmatize an incompetent doctor.[54] In the therapeutic configuration, Blacks and Whites alike find themselves de-psychologized. This

is the phenomenon that I wanted to isolate, taking advantage of the extreme experimental conditions in Saint-Denis. This is my own lactic acid ferment. The decisive innovation of ethnopsychiatric treatment stems, as I see it, from the re-creation inside a laboratory of a modus operandi whose effectiveness could not be measured in terms of notions of belief and representation. Divinities are not substances—no more so, in fact, than lactic acid ferment. They are all action.

Ethnographic literature abounds with descriptions of such goings-on, to be sure, but the savage whose portrait it draws remains a tinkering theoretician who slices the world into pieces in accordance with his own thinking. The primitive has indeed been saved, but only because we have attributed to him a theoretical mode of thought as close as possible to what we believed to be our own when we believed in beliefs, and therefore, in knowledge! Unfortunately, the science that is used for this defense owes everything to the theory of epistemologists, and nothing to the practice of charlatans.[55] Instead of comparing theories, let us compare practices—in the sense defined earlier in the shadow of factishes. No one can describe the coherence of a thought system better than an anthropologist, but nobody can do better than an ethnopsychiatrist in recreating the effectiveness of a gesture that here and now, in the Parisian suburbs, heals through the double artifice of the treatment configuration and an artificially induced affiliation.

Our list of specifications is slowly growing. These divinities exist; they are the subject of a positive discourse that is not mysterious in the least; they are not substances but a modus operandi; their effects can be observed on Whites as well as Blacks, under conditions that can be as artificial as one might wish, as long as everything is made to revolve around the gesture that heals.

We need to add one more feature before we can define these divinities: they cannot be mistaken for gods. The gods that save through "presence" are excellent vehicles for fabricating people, but poor agents for healing.[56] The subject constituted by the gods does indeed escape death, but that does not mean he is healed.[57] While the former subject of psychology could pull together the totality of his being within his own interiority, the one appearing here resembles a feuilleté, a multi-layered entity traversed by various vehicles each of which defines him in part without ever fully pinning him down. As we are beginning (or so I hope) to perceive, abandoning the difference between the interiorities of psychology and the exteriorities of epistemology is not the same as mixing everything up. When we lose the distinction between representations and facts, we by no means plunge into indifferentiation. If we follow the various vehicles, on the other hand, we can retrace distinctions other than the only two allowed by Modern scenography, continuing on to register many other contrasts.[58]

What characteristic hold do divinities have? What method of grasping, of passing? They construct those who seat them or fabricate them; this feature also belongs on the list of specifications. Without divinities, we die, as Jagannath understands in the middle of the courtyard, at the very moment he is sacralizing and desacralizing his family's saligram. More precisely, without a divinity we could not rid ourselves of other divinities who might threaten our existence. Each divinity thus appears as an anti-divinity. If we paid no attention to them, others might take their place. It may not be too far off to define them as quite a special sort of power relation. This is indeed how Jeanne Favret-Saada analyzed their meaning.[59] To find a model, we may need to turn to the complex sociology of apes, as described by the new primatologists,[60] or even to certain kinds of political relationships that Machiavelli analyzed, and that we find again in an almost pure state in international relations. There is nothing psychological in any of this. We find ourselves constantly threatened by forces that nevertheless have as a distinctive characteristic the fact that they can be overturned, or more precisely reversed, by a mere gesture. As Rilke wondered: "Perhaps all the dragons of our lives are princesses who are lonely waiting to see us once beautiful and brave."[61] Instead of forces continuously producing their effect, here are forces capable of brutally modifying their course, from dragons to princesses, from coaches to pumpkins, from saligrams to stones. The best one can do, under their sway, is to hold on a little longer, take a few more precautions, "watch out." Michel Serres has defined religion as "the opposite of negligence." There is indeed a significant religious element in the expectation that any second there might be some danger threatening us, because those to whom we owe our existence might not be able to come to our rescue.

Let us venture one more term to define these divinities at last. I propose to call them "fears," borrowing Tobie Nathan's lovely explanation for this word, which has the advantage of presuming neither essence nor person.[62] Fears do not need a personal subject, any more than the sentence "it is raining" does. Let us remember that the list of specifications I am trying to establish, a list that defines the mode of existence of these divinities, in no way entails the brute and stubborn existence of substance. Not only must fears abruptly reverse the maleficent or beneficent direction of their relationships, but they must also pass, or cause to pass. Their primary particularity, in fact, is that they never linger over the subject, whom they must completely ignore so that it may be safe a moment longer. They pass, they cross, they bounce over it; if they touch it, they do so by mistake, almost inadvertently; if they possess it, this is because they have mistaken their target. A series of lawless substitutions, they can transmute any being into any other being at any time. And that is why they elicit terror, and with good reason.[63]

I hesitate to reuse the term "transference," but in order to describe this movement, I should like to speak of transferences of fears: or better, *transfears*.[64] If the terms I am using are not grotesquely inadequate, to heal amounts to causing a fear that came out of nowhere to pass, to go elsewhere, anywhere, but especially to keep it from stopping: from attaching itself to the patient; from mistaking him for another and carrying him away; from substituting him, in its mad series of substitutions, for others who are never the same. To do that one must use trickery. Trickery inhabits this mode of being through and through. The fear must be tricked, at the price of complicated negotiations that are accounted for in terms borrowed from transactions, negotiations, or exchanges. Let us use the word charm instead, returning to the original meaning our language has lost. Charms allow one to use trickery against fear, according to the rather general formula, "if you can mistake me for another, you may take that other to be me," a necessary deception of which mythology provides hundreds of examples. Let us then imagine the provisional form of this quasi-subject of divinities, who would then replace the subject of psychology: surrounded by fears that can possess him by mistake, calling on counter-fears that are the object of continual scrutiny, he draws a rather loose envelope, patrolled by a multitude of charms, each of which deflects forces by using trickery. It is not a subject. It has neither interiority nor consciousness nor will. It is not yet fixated on Mommy-Daddy, as we have known since Deleuze's Anti-Oedipus. If it raves, it does so with the world, the cosmos, the socius, which it is exploring by a series of substitutions.[65]

We understand now that the invisibility of these fears does not result from a lack of existence. Nor does it come from the extra-terrestrial, extra-sensorial, super-natural, meta-psychical origin of so-called spirits: something that must be able to change meaning rapidly, by transforming itself all of a sudden, through a reversal of power relations, from good to evil or evil to good; something that must pass elsewhere on pain of possession and madness; something that must endlessly substitute one form for another, exploring, through free association, the combinations of the cosmos; something that can be deflected by the application of a tricky charm—all this cannot remain continuously, obstinately visible. Stubborn facts serve other purposes. Many other interactions, many other relationships, require continuity within being,[66] but not this one. Far from being invisible because their substance might include some mystery, transferences of fears remain invisible through the very simplicity of their felicity conditions. They are not occult, but have occulting lights. In other words, the mystery does not reside within them, but only in the twisting they are made to undergo, when optimization conditions belonging to others are applied to this particular vehicle, most often those of information transporters.[67] Ordered to transport forms and references, these words seem as weak as "abracadabras," bringing lovers of exoticism to despair. Charms, like angels, are very poor messengers.

By reformulating the metamorphosis of these invisible entities in my own inadequate language, I neither claim to have understood ethnopsychiatry, nor to have theorized it. Naturally, I was only interested in myself, or rather in those unfortunate Whites who are always being deprived of their anthropology by being locked into the modern destiny of antifetishism. By its very artifice, the ethnopsychiatric treatment session recreates laboratory conditions conducive to detecting invisible entities among us, in the suburbs of Paris. Time after time, it exposes clever therapeutic gestures and well-fabricated objects that seem to escape speech, but whose discourse, on the contrary, seems open to precise description, as long as we draw up the list of specifications of the entities mobilized, along with the optimization conditions of their involvement in action.[68] All I have done is choose the terms with care, so that they might pass from one end to the other of the ancient "great divide," sweeping aside one kind of phenomenon that neither object-less psychology nor subject-less epistemology seemed capable of accommodating. I am interested only in the questions that this reformulation allows me to raise, now that we have a broader and more symmetrical comparative base to draw on: since they have no more psychology than anyone else, what are the Whites' divinities? What invisibles are indispensable to the temporary and fragile construction of their envelopes and their quasi-subjects? How do they go about deflecting fears and transferring them elsewhere: by what charms, what tricks, and what mechanisms? Who are their ethnopsychiatrists?[69]

How to Understand an Action that has been Overtaken by Events

Why does Mafalda's father, in the last scene of this short comic strip (fig. 7), appear so terrified that he uses scissors to shred, compulsively, all the cigarettes remaining in his pack?[70] The reason is that Mafalda, incorrigible rascal, simply used the passive form to describe the innocuous behavior of her father.

"What are you doing?" she asks in the first scene.
"As you can see, I'm smoking," responds her father unwarily.

FIGURE 7.
"Le Club de Mafalda," no. 10 (Grenoble: Editions Glénat, 1986), 22.

"Oh," Mafalda remarks in passing, "I thought the cigarette was smoking you": panic.

Whereas he thought of himself as an untroubled father, comfortably seated in his armchair after a hard day at the office, his daughter saw him as an unbearable monster: a cigarette grabbing a man to have itself smoked in a big cloud of tar and nicotine; the father as an appendage, an instrument, an extension of the cigarette, the father become cigarette to the cigarette. Nothing more is needed to unleash a crisis: I foreswear smoking forevermore. To bind me to this promise, I reduce my entire pack to unsmokable stumps; I tear apart this idol that has enslaved me, into such minute fragments that it will never again be able to take hold of me, even if the craving, as we say, seizes me again.

Mafalda's amusing story has only the appearance of profundity. Moving from the first to the last scene, we basically pass from one extreme to another: at the start, the father believes himself given to an innocent vice that he has almost completely under control; at the end, he can extricate himself from his shackles only by pulverizing the cigarette, which so totally controls him that his daughter thought she had seen, in this hybrid, a cigarette smoking a man. In the two instances, both at the beginning and at the end, the reader continues to believe that we are talking about *control*. From the active form, "I smoke a cigarette," to the passive form, "you are smoked by a cigarette," nothing has changed other than the apportionment of master and instrument. The father alternates too drastically from one position to the other: too comfortable in the first image; too panicked in the last. What if the question rested instead on the absence of mastery, on the incapacity—either in the active or passive form—to define our attachments? How can we speak with precision of what the Greeks call "the middle voice", the verb form that is neither active nor passive?[71] The construal "factish" authorizes us to not take too seriously the ways in which subjects and objects are conventionally conjoined: that which sets into action never has the power of causation, whether it be a master subject or a causal object. That which is set into action never fails to transform the action, giving rise neither to the objectified tool nor to the reified subject. To think in terms of "factish" requires some getting used to, but once the initial surprise at such an outlandish form passes, one begins to regard those obsolete figures of object and subject, the made and maker, the acted upon and the actor, as more and more improbable.

Our vignette illustrates well their complete irrelevance: contrary to what Mafalda expresses in the fifth frame, the cigarette does not "smoke" her father, but without doubt, it is *making* the father smoke. This *faire-faire*, or "made to do," is so difficult to grasp that Mafalda's father thinks he escapes it by the two traditional routes: at the beginning, by thinking that he is capable of controlling his action: he acts and the cigarette does nothing; at the end—by thinking that he is completely controlled by the object—the cigarette acts and he does nothing. These two idioms, that of liberty and that of alienation, blinds us to the strange positioning of "factishes" capable of making one do things that no one, neither you nor they, can control. How does one become detoxified of this drug, mastery? What a surprising and almost contradictory question: *how does one emancipate oneself of the hard drug of emancipation?*

At this point, we need not be intimidated by the great battle between reactionaries and progressives. The former are categorically mistaken because they believe, on the pretext that detachment is not possible, that one must forever remain within the *same* attachments: a too convenient complacency that well justifies the indignation felt against those who want to leave the enslaved chained to masters of the past; a sufficient incitement to do battle against the injustice of fate and domination. Nonetheless, when the reactionaries mock the progressives by asserting that the liberation of the enslaved amounts to "changing the chains or the masters," the emancipator's indignation at these defeatist propositions is to be faulted: technically, the reactionaries are right and the progressives are wrong. When eulogizing liberty, the progressives forgot to specify, for those newly freed of their "bad" ties, the nature of the new ties with which, henceforth, they will be made to exist, the beings that will alienate them better. In speaking of liberty as an asymmetric term designating only the chains of the past, without referring to the bonds of the future, the progressives commit an error as flagrant as that of their ostensive opponents. Who is the sure assassin? The one who refuses to free the alienated from his mortifying ties, given that absolute liberty is a myth? Or the one who claims to de-alienate for good the subject, finally fully autonomous and master of himself, but without giving him the means to reestablish ties to those who are in a position to act upon him? Just a few years ago, the answer would have been easy: the first, without contest. Today, I admit without shame that I hesitate because my indignation requires that I now fight on two fronts, against the reactionaries and the progressives, the anti-moderns and the moderns.[72]

I am only interested and reassured by those who speak in terms of *substituting* one set of ties by another, and who, when they claim to unmake morbid ties, show me the new salutary ties, and this without ever looking to the subject as master of himself, now without an object. The terms liberation, emancipation, "laissez faire laissez-passer" must no longer assume automatic adherence by the "men (and women) of progress." Preceding the flag of Liberty, forever raised to guide the people, we would be well advised to carefully discriminate, among the attaching things themselves, those that will procure good and durable ties. From now on, the adherents of "factishes" — those adhered to and authorized by "factishes" — will refuse to equate, with Pavlovian reflex, emancipation with the highest good: liberty is not an ideal but a heritage to be sorted out.

Despite his iconoclastic gesture, Malfalda's father does not succeed, by "deconstructing" his pack of cigarettes, in obtaining his autonomy.

He succeeded only in passing from an extreme innocence to an extreme panic by way of four stages: he believed himself to be free; he becomes a slave in the eyes of his daughter; he panics; he liberates himself by breaking his chains. Basically, however, he has only shifted from one belief in his liberty, with cigarette, to another belief in his liberty, without. How would he have responded to the barbs of the pestering Mafalda if he had lived in the domain of factishes? In understanding the passive form "you are smoked by your cigarette," as an accurate approximation of the middle voice, he would have responded in the middle voice: "Yes, Mafalda, my daughter, I am effectively held by my cigarette, which makes me smoke it. There is nothing in this resembling a determining action, neither for it nor for me. I do not control it any more than it controls me. I am attached to it and, if I cannot hope for any kind of emancipation from it, then perhaps other attachments will come to substitute for this one, on condition that I don't panic and that you do not impose upon me — as a good critical sociologist of the left or a moralist of the right would — an ideal of detachment from which I would surely perish."[73] We can substitute one attachment with another, but we cannot move from a state of attachment to that of detachment. This is what a father should tell his daughter! To understand the activity of subjects, their emotions, their passions, we must turn our attention to that which attaches and activates them — an obvious proposition but one normally overlooked.

The single slogan "to live without a master" actually signifies two entirely different projects, depending on whether one lives under the umbrage of factishes or remains torn between objects and subjects. Does liberty consist *of living without a master, or without mastery?* The two projects are no more similar than *faire* and *faire-faire*, or "to do" and "to make do." The first project amounts to confounding the passage from one master to another with the passage from attachment to detachment. Behind the desire for emancipation — "neither God nor master" — lies the desire to substitute a good master for a bad one; most often, it entails the replacement of the institution with the "I/king," to adopt the expression of Pierre Legendre. Even if we accept that this merely represents a substitution and not a definitive severing of ties, freedom continues to consist of replacing one form of mastery by another. But when will we be able to untie ourselves from the ideal of mastery itself? When will we begin finally to taste the fruits of liberty, that is, "to live without a master," particularly without an I/king? This is the second project that gives an entirely different meaning to the same slogan. We had confused freedom as the exercise of command in the place of another commander, with freedom as life lived *without command altogether*. With factishes, the expression of freedom regains the path that the ideal of emancipation and detachment had transformed into an impasse: freedom becomes the right not to be deprived of ties that render existence possible, ties emptied of all ideals of determination, of a false theology of creation ex nihilo. If it is correct that we must replace the ancient opposition between the attached and detached with the substitution of good and bad attachments, this replacement would leave us only feeling stifled if it were not supplemented and completed by a second idea, i.e. the deliverance from mastery altogether: at all points of the network of attachments, the node is that of a make-make, not of a make nor of a made. That at least is a different project of emancipation, which is as vigorous as the former but much more credible because it obliges *us not to confuse living without control with living without attachments*.

In the eyes of those who have broken the *faire-faire*, who have sundered the factish, cultures of the past or at a distance seem profoundly incomprehensible. With the opposed notions of determination and freedom, of heteronomy and autonomy, how could we understand those forms of existence that claim, very simply, they could not exist without being continually intertwined with certain divinities or certain goods?[74] The notion of fetish or fetishism emerges precisely from the shock encounter between those who utilize the terms of necessity and freedom and those who know themselves to be fastened by numerous beings that make them exist.[75] Faced with the accusation cast by his daughter that he is totally dominated by his fetish, Mafalda's father has no other choice than fanatically to destroy his idol, guaranteeing that he again not succumb to a fatal attachment. His frenetic reaction proves that he is modern, but portends nothing positive about his ability to understand those ties that will make him and his daughter exist. We constantly deliberate to discern the meaning of those vague terms, the West and Modernity. We can define them simply enough: the Modern West has broken its fac- tishes and sees Others as bizarrely attached creatures, monsters as much in the grips of their beliefs and their passivity as the father viewed by his daughter Mafalda,[76] but it is the daughter who doesn't understand her father, the Westerner who doesn't understand the Other—rendered exotic by contrast with an ideal of detachment that would surely kill, if one were so mad as to actually apply it. The incapacity to recognize in oneself those attachments that enable one to act is taken as reason to believe oneself Western, and to imagine that the Others are not, but who are consequently entirely "Other," when in fact they differ only by what precisely attaches them. Instead of a great divide between Us and Them, between the detached and the attached, we would be better off introducing a number of small divides between those who are attached by one such set of particular entities and those attached by another such set of particular entities. The specific nature of the activating transfers makes all the difference, and not the astounding pretension of escaping all domination, whether by facts or by fetishes, by rationality or by ir- rationality. One gains alterity from attachments, not from the radical difference between the liberated and the alienated, the uprooted and the rooted, the mobile and the fixed.[77]

If we define politics as the progressive constitution of a common world, we can easily see how difficult it is to imagine a collective exis- tence, if all those who wished to participate were first asked to leave be- hind, in the outside vestibule, all the appurtenances and attachments which enabled them to exist. Westerners, as the masters of ceremony, manage not to apply the rule of abstention and detachment, which they apply to the "Others," to themselves. The Westerner's attachments are summed up by the two great accumulators of their distinctive tradition, Nature and Society: the reign of necessity and that of freedom. Usage of the term "globalization" permits one to believe that the common world will necessarily be an extension, in one form or another, of one of these two reigns. For the competing parties, the global framework of the discussion is not up for debate. Nothing proves, however, that the common world as the object of politics, or what Isabelle Stengers calls "cosmopolitics," resembles globalization.[78] Everything proves, on the contrary, that the two accumulators—the causal determination of Nature and the arbitrary arbitration of the Sovereign—no longer suffice to find closure to controversies concerning the progressive constitution of the common world. In a world that no longer moves from alienation to emancipation but from entanglement to even greater entanglement—no longer from the pre-modern to the modern but from the modern to the non-modern—the traditional division between determinations and liberations serves no useful purpose in defining a "globalization" whose complexity, for the moment, defies political understanding. Despite the automatic reaction of Mafalda's father, it is no longer a matter of abruptly passing from slavery to freedom by shattering idols, but of distinguishing those attachments that save from those that kill.[79]

Conclusion

I have given three different meanings to the "Modern cult of factish gods." As is customary in critical thought, I first reused the pejorative meanings of the words "fetish" and "worship." Once this is done, the Moderns can no longer appear to be without fetishes and worship, as they have boastingly—or despairingly—believed. They do have a fetish, the strangest one of all: they deny to the objects they fabricate the autonomy they have given them. They pretend they are not surpassed, outstripped by events. They want to keep their mastery, and they find its source within the human subject, the origin of action,[80] or else—in a brutal alternation that is now familiar to us—the Moderns, piqued at not being able to use human work as an explanation for action, want to kill the subject-source by drowning it in various effects of language, genes, texts, fields, subconsciousnesses, and causalities. "Since subjects can't have the complete mastery and freedom required by the Sartrian subject, then nobody will have them ever again!" they scream in anger. And they throw man himself onto the rubbish heap of their broken idols. The recuperators will then go to the dump in order to cobble together a subject endowed with rights. Existentialism, structuralism, the rights of man are the successive avatars of fetish worship by people who think they are very clever, because they think they have gotten rid of fetishes, beliefs, and naiveté for good, whereas no one has ever believed naively in fetishes: not even them!

I next assigned a second meaning to the expressions, one that restored value and power to the words "factish" and "worship." The hypothesis is much simpler, and in practice the Moderns have never abandoned it. The one who acts is not master of what he does or makes; others pass into action, and outstrip him. Still, this is not something that justifies drowning the subject in a sea of despair. Nowhere does there exist an acid capable of dissolving the subject. The latter receives its autonomy by giving the autonomy it does not possess to entities that come to life thanks to this conferral. The subject learns from mediation. It comes from factishes. It would die without them. If the expression seems difficult to grasp, let us compare it to the unlikely mechanism—with all its wheels and gears, contradictions, feedback loops, repairs, epicycles, dialectics, and contortions—of those marionettes and puppeteers we have seen tangled up in their strings, both visible and invisible, swimming in belief, guilt, insincerity, virtuality, and illusio. Seeking to make something simpler than factishes, the Moderns created something more complicated. Trying to make things more luminous, they made them more obscure. He who seeks to make an angel creates a man.

The Moderns have to worship factishes, mediations, and passages explicitly, since they have never had mastery over their own creations, and this is a good thing. The marionette image is justified here, as long as the puppeteer is asked a few questions.[81] She will tell you, as will everyone else—as will any creator and manipulator—that her marionettes dictate their behavior to her: that they make her act; that they express themselves through her; that she could never manipulate them or mechanize them. And yet she holds them, dominates and masters them. She will straightforwardly admit that she is slightly outstripped by what she controls. Now let us suppose that a second-level puppeteer comes along and ma-

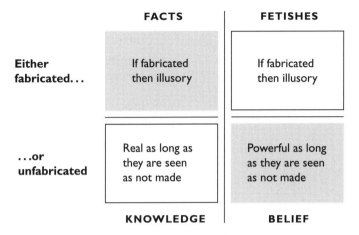

	FACTS	FETISHES
Either fabricated...	If fabricated then illusory	If fabricated then illusory
...or unfabricated	Real as long as they are seen as not made	Powerful as long as they are seen as not made
	KNOWLEDGE	BELIEF

FIGURE 8.

Knowledge and Belief are distinct because the theoretical argument radically separates the fabrication (top) from its result (bottom), and because it considers in Knowledge only the bottom left and in the Belief only in the top right (the two empty rectangles).

nipulates our artist. There is no lack of candidates: texts; language; the spirit of the times; the habitus; society; paradigms; epistemes; styles; any agent will do to hold our puppeteer as she holds her puppets. But these agents, no matter how powerful you make them, will be surpassed by the puppeteer, just as she is by her puppets. You will never do better; you will never hold on more tightly. Instead of a causal chain transmitting a force that actualizes a potential, realizes a possibility, you will never obtain more than a series of slight surpassings. Events: here is another name for the factishes and the worship they deserve.

But let us follow the whole chain; let us suppose that there is finally some master string-puller, finally some creator, an almighty being, an old-fashioned god, omniscient and omnipotent. This would change nothing. He would not be able to do anything more. A creature among creatures, he would also be slightly outstripped, surpassed by what he had fabricated; he would learn, from what he had created, about what he is made of: he would earn his autonomy through contact with his creatures just as we earn our bread by discovering, as we encounter other entities, something we did not know we were capable of a moment be-

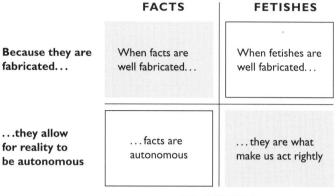

FIGURE 9.

In actual practice the fabrication is no longer denied, and the question shifts to the quality of the fabrication both for fetishes and for facts.

fore. Behind the pump of anti-fetishism hides a theology of creation, and it is quite a pitiful one, quite blasphemous. We imagine a creator god who would not be surpassed by what he makes, and who would master his creatures! Even when we deny his existence—especially when we deny it—this remains the model for action that I should like to usurp on behalf of humankind. Social constructivism is the poor man's creationism.[82] But there is no more creation by a source-god than there is construction by a source-human. Clerics, seeking to humble man's pride in his power to construct, may resort to the great trick of a creator god, but they are just as mistaken as the freedmen who claim to be cutting all ties and to be masters—without any masters above them—of what they fabricate, which lies beneath them. Really? Is an engineer master of his machine? Was Pasteur master of his lactic acid? Is a programmer master of his program, a creator, of his creation, an author, of his text? No one who has ever truly acted could utter such impieties. It is because God is a creature and because our creations possess, in relationship to ourselves, as much autonomy as we possess in relationship to God, that we can continue to use the words "construction" and "creation" without lying.[83]

If we have needed the complicated machinery of determinism, freedom, and grace for so long, it may be because we have not understood factishes. The slogan "neither god, nor master" should not be reserved for anarchists. It should also be engraved on the pedestals of the invisible, broken, and mended statues that allow action on all fronts. There may well be events, but no one is their master—especially not God.

Despite the vulgate, the term "ex nihilo" does not designate the primary matter animated by the demiurge, but the little threshold—the inevitable gap in *all mediated action*—that precisely renders demiurgic action impossible, since each event *exceeds* its conditions and hence exceeds its artificer. Whether we assert with Saint John, "at the beginning there was the made to speak, that is to say the Verb," or with Goethe, "at the beginning, there was the made to do, that is to say Action," in each case there is no creator in a position to dominate his creation drawn ex nihilo. As powerful as one might imagine a creator, he will never be capable of better controlling his creations than the puppeteer her puppets, a writer her notebooks, a cigarette its smoker, a speaker her language. He can make them do something but he cannot make them: to launch a cascade of irreversible events, yes, to be master of his tools, no. In believing that we were venerating the creator—God, humanity, subject or society—we chose, by a cruel deviation from theology, to idolize mastery as an ideal of detachment from everything that brought it into action. The expression ex nihilo doesn't signify that the artisan creates something out of nothing, but that the ensemble of prior conditions is never actually sufficient to determine action. What the word ex nihilo annihilates is the master's delusional pretension to mastery—and what is true for God is even truer for Man. There is only one perfume whose fragrance is agreeable to the Creator, that of surprise in beholding events that he does not control, but which he makes happen. The passage from nothingness to being or from being to nothingness has no part in the story—no more significant a part than the sudden swing from a careless freedom to a panicked fear of all forms of attachment for Mafalda's father. We would seriously misunderstand the redoubling inherent in *faire-faire* if we contented ourselves with stacking a second myth about creation on a first myth about creation. To use the word *faire-faire* signifies, on the contrary, that we wish to abandon completely the ideal of making, as well as its unfortunate "misdeeds."

"Should we allow djinns to be imported?" Tobie Nathan asks, and this is the third and last meaning I have given my title. The migrants wander—through the suburbs and even in Paris—with their divinities, but the worship of their factish gods is indeed fully of our time, since they live at once uprooted and re-rooted. In any case, this worship looks nothing like the cults of their past. The migrants have reconfigured the wisdom of the passage for us, stubbornly refusing to believe in their own gods, while we stubbornly believed that they were naively worshipping raw material and that we had left belief behind, entering into knowledge. Ethnopsychiatry may heal them, and so much the better; I am not in a position to judge. But the migrants heal us, at all events, and to that I can attest. They entertain entities in multiple, interesting, fragile states, without demanding that they stubbornly persist, or that they stem from our psychology. Thus they fray for us the difference between fabrication and reality, mastery and creation, constructivism and realism. They pass while talking endlessly, whereas we can only pass by hinting. They allow us to grasp more precisely our own sciences and technologies, fabrications that we might have thought they knew nothing about, or were dominated by. I find more accuracy in my lactic acid ferment if I shine the light of the Candomblé divinities on it. In the common world of comparative anthropology, lights cross paths. Differences are not there to be respected, neglected, or subsumed, but, as Whitehead said, to act as "lures for feelings, food for thought."

CHAPTER TWO

What Is Iconoclash?
Or Is There a World Beyond the Image Wars?

FIGURE 10.
Video still from "The Shroud," Alberto Giglio (color VHS, 2000).

This image comes from a video. What does it mean? Hooligans dressed in red, with helmets and axes, smash a reinforced window that protects a precious work of art. They madly hit the glass, which shatters in every direction while the crowd beneath them screams in horror, unable to stop the looting. Is it another sad case of vandalism captured by a video-surveillance camera? No, these are brave Italian firemen risking their lives, in the cathedral of Turin, to save the precious Shroud from a devastating fire. In their red uniforms and protective helmets, they use axes in an attempt to smash the heavily reinforced glass case that protects the venerable linen — not from vandalism — but from the mad passion of worshippers and pilgrims, who would have stopped at nothing to rend it in order to obtain priceless relics. The case is so well protected against worshippers that it cannot be brought to safety, away from the raging fire, without this violent act of glass breaking. With iconoclasm, one knows what the act of breaking represents, and what the motivations of apparent destruction are. For *iconoclash*, one does not know: one hesitates, one is troubled by an action for which there is no way to know, without further enquiry, whether it is destructive or constructive.

Why Do Images Trigger So Much Passion?

"Freud is perfectly right in insisting on the fact that we are dealing, in Egypt, with the first counter-religion in the history of humanity. It is here that, for the first time, the distinction has been made [by Akhenaton] that has triggered the hate of those excluded by it. It is since this distinction that hatred exists in the world and the only way to go beyond it is to go back to its origins."[1] What I propose here is to direct the attention of the reader to an archeology of hatred and fanaticism.[2]

Why? Because we are digging for the origin of an absolute — not a relative — distinction between truth and falsity, between a pure world, absolutely emptied of human-made intermediaries, and a disgusting world composed of impure but fascinating human-made mediators. "If only, some say, we could do without any image. How much better, purer, faster our access to God, Nature, Truth, and Science could be," to which other voices (or sometimes the same) answer: "Alas (or fortunately), we cannot do without images, intermediaries, mediators of all shapes and forms, because this is the only way to access God, Nature, Truth and Science." It is this quandary that we want to document, fathom and, maybe, overcome. In the strong summary that Marie-José Mondszain proposed of the Byzantine quarrel over images, "La vérité est image mais il n'y a pas d'image de la vérité" (Truth is image, but there is no image of truth).[3]

What has occurred to make images (and by image we mean any sign, work of art, inscription, or picture that acts as a mediation to access something else) the focus of so much passion? To the point that destroying them, erasing them, defacing them, has been taken as the ultimate touchstone, proving the validity of one's faith, one's science, one's critical acumen, and one's artistic creativity? To the point where being an iconoclast seems the highest virtue, the highest piety, in intellectual circles?

Furthermore, why is it that all those destroyers of images, those 'theoclasts', those iconoclasts, those 'ideoclasts' have also generated such a fabulous population of new images, fresh icons, rejuvenated mediators, greater flows of media, more powerful ideas, stronger idols? As if defacing some object would inevitably generate new faces, as if defacement and 'refacement' were necessarily coeval.[4] Even the tiny Buddha head that Heather Stoddard offered for our meditation, after having been smashed by the Red Guards during the Cultural Revolution, managed to take up a new sarcastic, cringing and painful face.[5]

And what has happened to explain that after every icono-crisis, infinite care is taken to reassemble the smashed statues, to save the fragments, to protect the debris? As if it were always necessary to apologize for the destruction of so much beauty, so much horror; as if one were suddenly uncertain about the role and cause of destruction that, before, seemed so urgent, so indispensable; as if the destroyer had suddenly realized that something else had been destroyed by mistake, something for which atonement was now overdue. Are not museums the temples in which sacrifices are made to apologize for so much destruction, as if we wanted suddenly to stop destroying and were beginning the indefinite cult of conserving, protecting, repairing?

This is what our exhibition attempted to do: the *capharnaüm* of heterogeneous objects that we assembled—broken, repaired, patched up, re-described—offered the visitors a meditation on the following questions:

Why have images attracted so much hatred?

Why do they always return again, no matter how strongly one
wants to get rid of them?

Why have the iconoclasts' hammers always seemed to strike side-
ways, destroying something else that seems, after the fact, to
matter immensely?

How is it possible to go beyond this cycle of fascination, repul-
sion, destruction, and atonement, which is generated by the
forbidden-image worship?

Iconoclasm

This essay, contrary to similar undertakings, is not iconoclastic: it is
about iconoclasm.[6] It attempts to suspend the urge to destroy images,
requires us to pause for a moment; to leave the hammer to rest. It prays
for an angel to come and arrest our arm, holding the knife ready to cut
the sacrificial lamb's throat. It is an attempt to turn around, envelop, and
embed the worship of image destruction: to give it a home, a site, a museum space, a place for meditation and surprise. Instead of iconoclasm
being the meta-language reigning as a master over all other languages,
it is the worship of iconoclasm itself that, in turn, is interrogated and
evaluated. From a resource, iconoclasm, is being turned into a topic. In
the words proposed by Miguel Tamen's beautiful title: I want readers to
become "friends of interpretable objects."[7]

I want to document, to expose, to do the anthropology of a certain
gesture, a certain movement of the hand. What does it mean to say of
some mediation—of some inscription—that it is human-made?

As is well-known from art historians and theologians, many sacred
icons that have been celebrated and worshipped are called *acheiropoiete*,
that is, not made by any human hand.[8] There are many instances of these
icons that have fallen from heaven without any intermediary: faces of
Christ; portraits of the Virgin; Veronica's veil. To show that a humble
human painter has made them would be to weaken their force, to sully
their origin, to desecrate them. Thus, to add the hand to the pictures is

tantamount to spoiling them, criticizing them. The same is true of religion in general. If you say it is man-made you nullify the transcendence of the divinities, you empty the claims of a salvation from above.

More generally, the critical mind is one that shows the hands of humans at work everywhere, so as to slaughter the sanctity of religion, the belief in fetishes, the worship of transcendent heaven-sent icons, the strength of ideologies. The more the human hand can be seen as having worked on an image, the weaker is the image's claim to offer truth. Since Antiquity, critics have never tired of denouncing the devious plots of humans who try to make others believe in non-existing fetishes. The trick to uncovering the trick is always to show the lowly origin of the work, the manipulator, the counterfeiter, or the fraud behind the scenes, caught red-handed.

The same is true of science. There too, objectivity is supposed to be *acheiropoiete*, not made by human hand. If you show the hand at work in the human fabric of science, you are accused of sullying the sanctity of objectivity, of ruining its transcendence, of forbidding any claim to truth, of putting to the torch the only source of enlightenment we may have.[9] I treat as iconoclasts those who speak of the humans at work—scientists in their laboratories—behind or beneath the images that generate scientific objectivity. I have also been held by this paradoxical *iconoclash*: the new reverence for the images of science is taken to be their destruction. The only way to defend science against the accusation of fabrication, to avoid the label of "socially constructed," is apparently to insist that no human hand has ever touched the image it has produced.[10] So, in the two cases of religion and science, when the hand is shown at work, it is always a hand with a hammer or with a torch: always a critical; a destructive hand.

What if hands were actually indispensable to reaching truth, to producing objectivity, to fabricating divinities? What would happen if, when saying that some image is human-made, you were increasing instead of decreasing its claim to truth? That would be the closure of the critical mood, the end of anti-fetishism. We could say, contrary to the critical urge, that the more humans there are, the more human-work is shown: the better is their grasp of reality, of sanctity, of worship. The more images, mediations, intermediaries, and icons are multiplied and overtly fabricated—explicitly and publicly constructed—the more respect we have for their capacities to welcome, gather, and recollect truth and sanctity (*religere* is one of the several etymologies for the word religion). As Mick Taussig has so beautifully shown, the more you reveal the tricks necessary to invite the gods to the ceremony during the initiation, the more certain the divinities are present.[11] Far from despoiling access to transcendent beings, the revelation of human toil, of the tricks, reinforce the quality of this access.[12]

Thus, we can define an *iconoclash* as that which happens when there is uncertainty about the exact role of the hand at work in the production of a mediator. Is it a hand with a hammer ready to expose, denounce, debunk, show up, disappoint, disenchant, dispel one's illusions, or let the air out? Or is it, on the contrary, a cautious and careful hand, with palm turned as if to catch, elicit, educe, welcome, generate, entertain, maintain, or collect truth and sanctity?

Then, of course, the second commandment can no longer be obeyed: "You shall not make for yourself an idol in the form of anything in heaven above or on earth beneath or in the waters below." No need to try to fudge the intention and tension of this essay: it is about the second commandment. Are we sure we have understood it correctly? Have we not made a long and terrifying mistake about its meaning? How can we reconcile this request for a totally aniconic society, religion, and science with the fabulous proliferation of images that characterizes our media-filled cultures?

If images are so dangerous, why do we have so many of them? If they are innocent, why do they trigger so many and such enduring passions? Such is the enigma, the hesitation, the visual puzzle, the *iconoclash* that I wish to deploy under the eyes of the reader.

Religion, Science, and Art:
Three Different Patterns of Image-Making

To begin to answer these questions it is important to compare three sources of *iconoclashes*: religion, science, and contemporary art. This is not to offer a religious pilgrimage, nor to learn about science, and still less to review contemporary art, but to compare for each of these three sets of images different patterns of belief, rage, enthusiasm, admiration, diffidence, fascination, and suspicion. What interests me in proposing this unusual combination is the set of interferences these different patterns offer.

Icons and Idols

Have religious icons not been emptied by aesthetic judgment, absorbed by art history, made routine by conventional piety, to the point of being dead forever? On the contrary, it is enough to remember the reactions to the destructions of the Bamiyan Buddhas by the Taliban in Afghanistan, to realize that religious images are still the ones that attract the fiercest passions.[13] From Akhenaton's 'theo-clast' onwards, destroying monasteries, churches, and mosques, and burning fetishes and idols in huge bonfires, is still a daily occupation for huge masses of the world, exactly as in the time of what Assmann calls the "mosaic distinction."[14] "Break down their altars, smash their sacred stones and burn their sacred poles" (Exodus 34, 13): the instruction to burn the idols is as present, as burning, as impetuous, as subterraneous as the ever-threatening lava flows along the Etna. Even in the hilarious case of the destruction, in the summer of 2001, of the "Mandarom"—a hideous, gigantic statue erected by a sect in the South of France—which the believers have compared to the demise of the Afghan Buddhas.

Of course, idol smashing is in no way limited to religious minds. Which critic does not believe that her ultimate duty, her most burning commitment, is to destroy the totem poles, expose ideologies, and disabuse the idolaters? As many people have remarked, 99 percent of those who were scandalized by the Taliban gesture of vandalism descended from ancestors who had smashed the most precious icons of some other people—or they had themselves participated in some deed of deconstruction.[15]

What has been most violent? The religious urge to destroy idols to bring humanity to the right cult of the true God, or the anti-religious urge to destroy the sacred icons and bring humanity to its true senses? An *iconoclash* indeed, since, if they are nothing, no one knows whether those idols can be smashed without any consequences ("They are mere stones," said Mollah Omar,[16] in the same fashion as the Byzantine and later Lutheran iconoclasts) or whether they have to be destroyed because they are so powerful, so ominous ("If they are so vacuous why do you take up on them?" "Your idol is my icon").[17]

Scientific Inscriptions

But what is one to make of scientific images in this context? Surely, these offer cold, unmediated, objective representations of the world, and thus cannot trigger the same passion and frenzy as the religious pictures. Contrary to the religious ones, they simply describe the world in a way that can be proven true or false. Precisely because they are cool, fresh, verifiable, and largely undisputed, they are the objects of a rare and almost universal agreement. So the pattern of confidence, belief, rejection, and spite is entirely different for them than the one generated by idols or icons. This is why they offer different sorts of *iconoclashes*.

To begin with, for most people, they are not even images, but the world itself. There is nothing to say about them but to learn their message. To call them image, inscription, or representation, to have them exposed in an exhibition side by side with religious icons, is already an iconoclastic gesture. If those are mere representations of galaxies, atoms, light, or genes, then one could say indignantly, "they are not real, they have been fabricated." And yet, as will be made apparent here, it slowly becomes clearer that without huge and costly instruments, large groups of scientists, vast amounts of money, and long training, nothing would be visible in such images. It is because of so many mediations that they are able to be so objectively true.[18]

Here is another *iconoclash*, exactly opposite of the one raised by the worship of religious image-destruction: the more instruments, the more mediation, the better the grasp of reality.[19] If there is a domain where the second commandment cannot be applied, it is the one ruled by those who shape objects, maps, and diagrams "in the form of anything in heaven above or on earth beneath or in the waters below." So the pattern of interference may allow us to rejuvenate our understanding of image making: the more human-made images are generated, the more objectivity will be collected. In science, there is no such a thing as mere representation.

Contemporary Art

Then why link religious and scientific mediations to contemporary art? Because here at least there is no question that paintings, installations, happenings, events, and museums are human-made. The hand at work is visible everywhere. No *acheiropoiete* icon is expected from this maelstrom of movements, artists, promoters, buyers and sellers, critics and dissidents. On the contrary, the most extreme claims have been made in favor of an individual, human-based creativity, with no access to truth or to the divinities. Down with transcendence![20]

Nowhere else but in contemporary art has a better laboratory been set up for trying out and testing the resistance of every item comprising the cult of image, picture, beauty, media, or genius. Nowhere else have so many paradoxical effects been carried out on the public to complicate their reactions to images.[21] Nowhere else have so many set-ups been invented to slow down, modify, perturb, and lose the naive gaze and the scopic regime of the *amateur d'art*.[22] Everything has been slowly experimented against and smashed to pieces, from mimetic representation, through image making, canvas, color, and artwork, all the way to the artist herself, her signature, the role of museums, of the patrons, of critics—not to forget the Philistines, ridiculed to death.

Everyone and every detail of what art is and what an icon is—an idol, a sight, a gaze — has been thrown into the pot to be cooked and burnt up in the past century of what used to be called modernist art.[23] A Last Judgment has been passed: all our ways of producing representation of any sort have been found wanting. All that remains are generations of iconoclasts smashing each other's faces and works: a fabulous large-scale experiment in nihilism; a maniacal joy in self-destruction; a hilarious sacrilege; a sort of deleterious an-iconic inferno.[24]

As one might expect, here is another *iconoclash*: so much defacement and so much "re-facement."[25] Out of this obsessive experiment to avoid the power of traditional image making, a fabulous source of new images, new media, and new works of art has been found: new set-ups to multiply the possibilities of vision. The more art has become a synonym for the destruction of art, the more art has been produced, evaluated, talked about, bought and sold, and yes, worshipped. New images have been produced, so powerful that they have become impossible to buy, touch, burn, repair, or even transport, thus generating even more *iconoclashes*: a sort of 'creative destruction' that Schumpeter had not anticipated.[26]

A Reshuffling of Confidence and Diffidence Towards Image

We are considering three different patterns of image rejection and image construction, of image confidence and image diffidence. My bet is that interference between the three should move us beyond the image wars, beyond the *bildersturm*. I have not considered religious images in order to have them again subjected to irony or destruction, nor to again present them to be worshipped. I consider them to resonate with scientific images, demonstrating which ways they are powerful and what sort of invisibility both types of images have been able to produce.[27]

Scientific images have not been considered here to instruct or enlighten the public in some pedagogical way, but to show how they are generated and how they connect to one another, to which sort of iconoclasm they have been subjected, and what peculiar type of invisible world they generate.[28]

As to contemporary art pieces, they are not being considered here to compose a critique, but to draw the conclusions of this experiment on the limits and virtues of representation that has been going on in so many media, and through so many bold innovative enterprises.[29]

In effect, I am trying to consider, for recent iconoclastic art, a sort of idol-chamber, similar to the ones made by protestant desecrators when they tore the images away from cult, turning them into objects of horror and derision, before they became the first kernels of the art museum and aesthetic appreciation.[30] This is a little twist to be sure, and more than a little ironic, but much welcome.

The routine patterns of respect, wonder, diffidence, worship, and confidence, which usually distinguish religious, scientific and artistic mediations should be, more than slightly redistributed throughout this essay.

Which Object to Select?

As should be clear by now, this essay is not a philosophical argument, but a cabinet of curiosities, assembled by a "friend of interpretable objects" to fathom the source of the fanaticism, hatred, and nihilism generated by the image in Western tradition. It is not a small project, but since I am not totally mad, I have not tried to cover the whole question of image worship and destruction from Akhenaton to 9/11. This is not an encyclopedic undertaking. On the contrary, I have very selectively chosen only those sites, objects, and situations where there is an ambiguity, a hesitation, an *iconoclash* on how to interpret image-making and image-breaking. I am interested in representing the state of mind of those who have broken fetishes—what I prefer to call "factishes"[31]—and who have entered into what Assmann names "counter-religion."

An Impossible Double Bind

How can they stand living with the broken pieces of what, until they came along, had been the only way to produce, to collect, to welcome the divinities? How startled they should be when they look at their hands, which are no longer able to complete the tasks they had succeeded in doing for eons: namely, to be busy at work and nonetheless to generate objects which are not of their own making? Now they have to choose between two contradictory demands: is this made by your own hands, in which case it is worthless; or is this objective, true, transcendent, in which case you cannot possibly have made it? Either God is doing everything and humans are doing nothing, or the humans are doing all the work and God is nothing: too much or too little when the fetishes are gone.

Fetishes have to be made, of course. Human hands cannot stop toiling: producing images, pictures, inscriptions of all sorts; to still generate, welcome, and collect objectivity, beauty, and divinities, exactly as in the now-forbidden, repressed, obliterated old days. How could one not become a fanatic since gods, truths, and sanctity have to be made and there is no longer any legitimate way of making them? How can one live with this double bind without becoming mad? Have we become mad? Is there a cure to this folly?

Let us contemplate for a moment the tension created by this double bind, which may explain a lot of the archeology of fanaticism. The idol smasher, the mediator-breaker is left with only two polar opposites: either he (I guess it is fair to put it in the masculine) is in full command of his hands—but then what he has produced is simply the mere consequence of his own force and weakness projected into matter, since he is unable to produce more output than his input; in which case, there is no other way for him but to alternate between hubris and despair, depending on whether he emphasizes his infinite creative power or his absurdly limited forces—or he is in the hands of a transcendent, unmade divinity who has created him out of nothing and produces truth and sanctity in the *acheiropoietic* way. In the same way that he, the human fabricator, alternates between hubris and despair, He—the Creator—will alternate wildly between omnipotence and non-existence, depending on whether or not His presence can be shown and His efficacy proven. What used to be synonymous, "I make," and "I am not in command of what I make," has become a radical contradiction, "Either you make or you are made."[32]

This brutal alternation between being in command as a powerful (powerless) human creator, or being in the hand of an omnipotent (powerless) Creator, is already bad enough, but worse, what really knots the double bind and forces the strait-jacketed human into extreme frenzy, is that there is no way to stop the proliferation of mediators, inscriptions, objects, icons, idols, image, picture, and signs, in spite of their interdiction. No matter how adamant one is about breaking fetishes and forbidding oneself image-worship, temples will be built, sacrifices will be made, instruments will be deployed, scriptures will be carefully written down, manuscripts will be copied, incense will be burned, and thousands of gestures will have to be invented for recollecting truth, objectivity, and sanctity.[33]

The second commandment is all the more terrifying since there is no way to obey it. The only thing you can do to pretend you observe it is to deny the work of your own hands, to repress the action ever present in the making, fabrication, construction, and production of images, to erase the writing at the same time you are writing it, to slap your hands at the same time they are manufacturing. With no hand, what will you do? With no image, to what truth will you have access? With no instrument, what science will instruct you?

Can we measure the misery endured by those who have to produce images and are forbidden to confess they are making them? Worse, either they will have to say that the demiurge is doing all the work—writing the sacred scriptures directly, inventing the rituals, ordering the law, assembling the crowds—or else, if the work of the faithful is revealed, we will be forced to denounce those texts as "mere" fabrications, those rituals as make-believe, their making as nothing but making up, their constructions as a sham, their objectivity as socially constructed, their laws as simply human, too human.[34]

So the idol-smasher is doubly mad: not only has he deprived himself of the secret to produce transcendent objects, but he continues producing them even though this production has become absolutely forbidden, with no way to be registered. Not only does he hesitate between infinite power and infinite weakness, infinite creative freedom and infinite dependence in the hand of his Creator, but also he constantly alternates between the denial of the mediators and their necessary presence. This is enough to render one mad, enough at least to produce more than one *iconoclash*.

Freud, in his strange nightmare about Moses, has offered to explain a similar madness—the invention of counter-religion—a most bizarre legend, that of the murder of the selfish, overpowering father by the primitive horde of his jealous sons.[35] But the tradition offers another, more revealing legend, where it is not the father that is killed, but the father's livelihood that is smashed to pieces by his over enterprising son.[36] At the age of six, Abraham is said to have destroyed the idol-shop of his father, Terah, with which he had been temporarily entrusted. The story is recounted in the Midrash Rabbah:

> Rabbi Hiya the son of Rabbi Ada said that Terah [Abraham's father] was an idol worshipper. One day Terach had to leave the store [in which he sold idols]. He left Abraham to manage the store in his absence. A man came and wanted to buy an idol. Abraham asked him "How old are you?" And he responded "Fifty or sixty years old." Abraham then said, "Pitiful is the man who is sixty and worships idols that are only a day old." So the man left in embarrassment. Once, came a woman with an offering of fine flour. She said to him [Abraham] "Here, take it and bring it before" [the idols]. Abraham stood up, took a stick, broke all the idols, and put the stick back in the hands of the biggest idol among them. When his father returned he asked "Who did this to them?" Abraham answered, "I will not deny you the truth. A woman came with an offering of fine flour and asked me to bring it before them. So I brought it before them, and each said, 'I shall eat first.' Then the biggest one stood among them, he took a stick in his hand and broke them all." So Terach said to him, "Why do you mock me? Do these [idols] know anything [to speak and move]?" And Abraham replied, "Won't your ears hear what your mouth speaks?"[37]

What a good *iconoclash*! To this day no one understands the ambiguous response of the father to the son's question: "Why does your ear not listen to what your mouth says?" Is the son shaming his father for his idol worship, or is it, on the contrary, the father who is shaming his son for not understanding what idols can do?[38] If you start to break the idols, my son, with what mediations will you welcome, collect, access, assemble, and gather your divinities? Are you sure you understand the dictates of your God? What sort of folly are you going to enter if you begin to believe, that I, your father, naively believe in those idols I have made with my own hands, cooked in my own oven, sculpted with my own tools? Do you really believe I ignore their origin? Do you really believe that this lowly origin weakens their claims to reality? Is your critical mind so very naive?

This legendary dispute can be seen everywhere in more abstract terms, whenever a productive mediation is smashed to pieces and replaced by the question: Is this made or is this real? You have to choose![39] What has rendered constructivism impossible in the Western tradition? It is a tradition that, on the other hand, has constructed and deconstructed so much, but without being able to confess how it managed to do it. If westerners had really believed they had to choose between construction and reality (if they had been consistently modern), they would never have had religion, art, science, or politics. Mediations are necessary everywhere. If you forbid them, you may become mad, fanatic, but there is no way to obey the command and choose between the two-polar opposites: either it is made or it is real. That is a structural impossibility, an impasse, a double bind, a frenzy. It is as impossible as to request a Bunraku player to have to choose, from now on, either to show his puppet or to show himself on the stage.

To Increase the Cost of Criticism

For my part, I have selected items that reveal this double bind and the fanaticism it triggers.[40] It is as if the critical mind could not overcome the original breaking of "factishes" in order to realize how much it had lost in forcing the fabricator into this impossible choice, between human construction and access to truth and objectivity. Suspicion has rendered us dumb. It is as if the hammer of the critique had rebounded and struck senseless the critic's head.

This is also a revision of the critical spirit, a pause in the critique, a meditation on the urge for debunking, for the too quick attribution of the naive belief in others.[41] The devotees are not dumb.[42] It is not that critique is no longer needed, but rather that it has, of late, become too cheap.

One could say, with more than a little dose of irony, that there has been a sort of miniaturization of critical efforts: what in the past centuries required the formidable effort of a Marx, a Nietzsche, a Benjamin, has become accessible for nothing, much like the supercomputers of the 1950s, which used to fill large halls and expend a vast amount of electricity and heat, and are now accessible for a dime, and no bigger than a fingernail. You can now have your Baudrillard's or your Bourdieu's disillusion for a song, your Derridian deconstruction for a nickel. Conspiracy theory costs nothing to produce; disbelief is easy, debunking what is learned in 101 classes in critical theory. As the recent advertisement of a Hollywood film proclaimed, "Every one is suspect . . . everyone is for sale . . . and nothing is true!"

I wish to make critique more difficult, to increase its cost, by adding another layer to it, another *iconoclash*: what if the critique had been uncritical enough to the point of making invisible the necessity of mediation? What is the western underbelly, modernism's hidden spring that makes its machinery tick? Again, what if we had misunderstood the second commandment? What if Moses had been forced to tone it down for the narrow bandwidth of his people?

A Rough Classification of the Iconoclastic Gestures

It might be useful to produce a classification of the *iconoclashes* considered here. It is of course impossible to propose a standardized, agreed-upon typology for such a complex and elusive phenomenon.

It would even seem to run counter to the spirit of the book. As I have claimed, somewhat boldly: are we not after a re-description of iconophilia and iconoclasm in order to produce even more uncertainty about which kind of image worship/image smashing one is faced with? How could we neatly pull them apart? Yet it might be useful briefly to present the five types of iconoclastic gestures reviewed in this show, for no better reason than to gauge the extent of the ambiguity triggered by the visual puzzles we have been looking for.

The principle behind this admittedly rough classification is to look at:

The inner goals of the icon smashers.

The roles they give to the destroyed images.

The effects this destruction has on those who cherished those
   images.

How this reaction is interpreted by the iconoclasts.

And, finally, the effects of destruction on the destroyer's own
   feelings.

This list is rudimentary but sturdy enough, I think, to guide one through the many examples assembled here.

The A People are Against All Images

The first type—I give them letters to avoid loaded terminology—is made up of those who want to free the believers, those they deem to be believers, of their false attachments to idols of all sorts and shapes. Idols, the fragments of which are now lying on the ground, were nothing but obstacles in the path to higher virtues. They had to be destroyed. They triggered too much indignation and hatred in the hearts of the courageous image breakers. Living with them was unbearable.[43]

What distinguishes the As from all other types of iconoclasts is that they believe it is not only necessary, but also possible, to dispose entirely of intermediaries for access to truth, objectivity, and sanctity. Without those obstacles, they think one will at last have smoother, faster, and more direct access to the real thing, which is the only object worthy of respect and worship. Images do not even provide preparation, a reflection, or an inkling of the original: they forbid any access to the original. Between images and symbols, you have to choose or be damned.

Type A is thus the pure form of classical iconoclasm, recognizable in the formalist's rejection of imagination, drawing, and models,[44] as well as in the many Byzantine, Lutheran, revolutionary movements of idol smashers, and in the horrifying "excesses" of the Cultural Revolution.[45] Purification is their goal. The world, for A people, would be a much better place, much cleaner, much more enlightened, if only one could get rid of all mediations: if one could jump directly into contact with the original, the ideas, the true God.

One of the problems with the As is that they have to believe that the others—the poor guys whose cherished icons have been accused of being impious idols—believe naively in them. Such an assumption entails that, when the philistines react with screams of horror to pillage and plunder, this does not stop the As. On the contrary, it proves how right they were.[46] The intensity of the horror of the idolaters is the best proof that those poor naive believers had invested too much in their stones, which are essentially nothing. Armed with the notion of naive belief, the freedom-fighters constantly misconstrue the indignation of those they scandalize, in order to effect an abject attachment to things they should destroy even more radically.

But the deepest problem of the As is that no one knows whether they are not Bs!

The B People are Against Freeze-Frame, not Against Images

The Bs too are idol smashers. They also wreak havoc on images, break down customs and habits, scandalize the worshippers, and trigger the horrified screams of Blasphemer! Infidel! Sacrilege! Profanity! The difference between the As and the Bs is that the latter do not believe it possible nor necessary to get rid of images. What they fight is freeze-framing, that is, extracting an image out of the flow, becoming fascinated by it, as if it were sufficient, as if all movement had stopped.

What they are after is not a world free of images, purified of all the obstacles, rid of all mediators, but on the contrary, a world filled with active images, with moving mediators. They do not want the image production to stop forever—as the As will have it—they want it to resume as fast and as fresh as possible.

For them, iconophilia does not mean the exclusive and obsessive attention to image, because they can stand fixed images no more than the As. Iconophilia means moving from one image to the next. They know "truth is image but there is no image of truth." For them, the only way to access truth, objectivity, and sanctity is to move fast from one image to another, not to dream the impossible dream of jumping to a nonexisting original. Contrary to Plato's resemblance chain, they don't even try to move from the copy to the prototype. They are, as the old iconophilic Byzantine used to say, "economic,"[47] a word meaning, for the Byzantines, a long and carefully managed flow of images in religion, politics, and art, but not the sense it now has: the world of goods.

Whereas the As believe that those who hold to images are iconophilic, and that the courageous minds who break away from the fascination with images are iconoclastic, the Bs define as iconophilic those who do not cling to one image in particular but who are able to move from one to the other. For them, iconoclasts are either those who absurdly want to get rid of all images, or those who remain in the fascinated contemplation of one isolated image, freeze-framed.

Prototypical examples of Bs might include Jesus chasing the merchants out of the Temple, Bach shocking the dull music out of the Leipzig congregation's ears,[48] Malevich painting the black square to access the cosmic forces that had remained hidden in classical representative painting,[49] or the Tibetan sage extinguishing the butt of a cigarette on a Buddha's head in order to show its illusory character.[50] The damage done to icons is, to them, always a charitable injunction to redirect their attention towards other, newer, fresher, more sacred images: not to do without image.

But of course many *iconoclashes* come from the fact that no worshipper can be sure when his or her preferred icon/idols will be smashed to the ground, or whether an A or a B does the ominous deed. Are we requested, they wonder, to go without any mediation at all and try out direct connections with God and objectivity? Are we simply invited to change the vehicle we have used so far for worship? Are we spurred into a renewed sense of adoration and asked to resume our work of image-building anew? Think of the long hesitation of those waiting at the foot of Mount Sinai for Moses to return: what have we been asked to do? It is so easy to be mistaken and to begin molding the golden calf.[51]

Are neither the As nor the Bs sure of how to read the reactions of those whose icon/idols are being burnt? Are they furious at being without their cherished idols, much like toddlers suddenly deprived of their transitional object? Are they ashamed of being falsely accused of naively believing in non-existing things? Are they horrified at being so forcefully requested to renew their adhesion to their cherished tradition, which they had let fall into disrepute and mere custom? Neither the As nor the Bs can decide, from the screeching noise made by their opponents, what sort of prophets they are: are they prophets who claim to get rid of all images, or the ones who, "economically," want to let the cascade of images move again to resume the work of salvation?

But this is not the end of our hesitation, our ambiguity, our *iconoclash*. As and Bs could, after all, simply be Cs in disguise.

The C People are not Against Images, Except Those of Their Opponents

The Cs are also after debunking, disenchantment, idol-breaking. They too leave in their trail plunder, wreckage, horrified screams, scandals, abomination, desecration, shame, and profanation of all sorts. But contrary to the As and to the Bs, they have nothing against images in general: they are only against the image to which their opponents cling most forcefully.

This is the well-known mechanism of provocation by which, in order to destroy someone as fast and as efficiently as possible, it is enough to attack what is most cherished, what has become the repository of all the symbolic treasures of one people.[52] Flag-burning, the slashing of paintings, and hostage-taking are typical examples. Tell me what you hold to be most dear and I will wreck it so as to kill you faster. It is the minmax strategy so characteristic of terrorist threats: the maximum damage for the minimum investment. Box cutters and plane tickets against the United States of America.

The search for the suitable object to attract destruction and hatred is reciprocal: "Before you wanted to attack my flag, I did not know I cherished it so much, but now I do."[53] So the provocateurs and those they provoke are playing cat and mouse, the first looking for what triggers indignation faster, the others looking eagerly for what will trigger indignation most fiercely.[54] During this search, all recognize the image in question as a mere token; it counts for nothing but an occasion that allows the scandal to unfold.[55] If it were not for the conflict, everyone in the two camps would be perfectly happy to confess that it is not the object that is disputed: it is just a stake for something entirely different.[56] So for the Cs, the image itself is not in question at all, they have nothing against it (unlike the As) or for it (as in the case of the Bs). The image is simply worthless; it is worthless but attacked, thus defended, thus attacked, and on to infinity.

What is so terrible for idol smashers is that there is no way to decide for good whether they are As, Bs, or Cs. Maybe they have entirely misunderstood their calling; maybe they are misconstruing the screams of horror of those they call philistines, who witness their idols smashed to the ground. They see themselves as prophets, but it may be they are mere *agents provocateurs.* They see themselves as freeing the poor wretched souls from their imprisonment by monstrous things, but what if they were, on the contrary, scandalmongers looking for ways to shame their opponents most efficiently?

What would happen to me if, in criticizing the critics, I was simply trying to create another scandal? What if this essay, in its pretension to re-describe iconoclasm, was nothing but another boring iconoclastic gesture, another provocation, the mere repetition of the endless gesture of the intelligentsia's most cherished treasures? We don't know for sure.

The D People are Breaking Images Unwittingly

There is another kind of icon smasher, a most devious case, those who could be called the "innocent vandals." As is well known, vandalism is a term of spite invented to describe those who destroy not so much out of a hatred of images, but out of ignorance, a lust for profit, and sheer passion and lunacy.[57]

Of course, the label can be used to describe the action of the As, Bs, and Cs as well. They all can be accused of vandalism by those who don't know if they are innocent believers, furious at being accused of naiveté, philistines awakened from their dogmatic sleep by prophetic calls, or lovers of scandal delighted at being the butt of criticism, and thus able to demonstrate the strength and self-righteousness of their indignation.

But the innocent vandals are different from the normal, "bad" vandals: they had absolutely no idea that they were destroying anything. On the contrary, they were cherishing images and protecting them from destruction, and yet they are accused later of having profaned and destroyed them.[58] They are, so to speak, iconoclasts in retrospect. The typical example is that of the restaurateurs who are accused by some of "killing with kindness."[59] The field of architecture is especially filled with those "innocents" who, when they build, have to destroy, when their buildings are accused of being nothing but vandalism.[60] Their hearts are filled with the love of images—so they are different from all the other cases—and yet they trigger the very same curses of "profanation," "sacrilege," and "desecration" as all the others.

Life is so difficult: by restoring works of art, beautifying cities, rebuilding archeological sites, they have destroyed them, their opponents say, to the point that they appear as the worst iconoclasts, or at least the most perverse ones. But other examples can be found, like those museum curators who keep the beautiful New Guinean mallagans—even though they have become worthless since, in the eyes of their makers, they should be destroyed after three days[61]—or those African objects that have been carefully made to rot on the ground, and which are carefully saved by art dealers and thus rendered powerless—in the eyes of their makers.[62] The apprentice sorcerer is not a really wicked sorcerer, but one who becomes wicked out of his or her own innocence, ignorance, and carelessness.

And here again the As—as well as the Bs and Cs—can be accused of being Ds: that is, accused of aiming at the wrong target, of forgetting to take into account the side effects, or the far reaching consequences of their acts of destruction. "You believe you freed people from idolatry, but you have simply deprived them of the means to worship"; "You believe you are a prophet renewing the cult of images with fresher images, but you are nothing but a scandal-monger thirsty for blood." These and similar accusations are frequently leveled in revolutionary circles, accusing one another of being constantly on the wrong foot, of being, *horresco referens*, reactionary. What if we had killed the wrong people, smashed down the wrong idols? Worse, what if we had sacrificed idols for the cult of an even bloodier, bigger, and more monstrous Baal?

The E People are Simply the People: They Mock Iconolasts and Iconophiles

To be complete, one should add the Es, who doubt the idol breakers as much as the icon worshippers. They are diffident to any sharp distinctions between the two poles; they exercise their devastating irony against all mediators: not that they want to get rid of them, but because they are so conscious of their fragility. They love to show irreverence and disrespect; they crave for jeers and mockery; they claim an absolute right to blasphemy in a fierce, Rabelaisian way;[63] they show the necessity of insolence, the importance of what the Romans called *pasquinades*, which is so important for a healthy sense of civil liberty, the indispensable dose of what Peter Sloterdijk has called kynicism (by opposition to the typically iconoclastic cynicism).[64]

There is a right not to believe, and the even more important right not to be accused of believing naively in something. There may be no such a thing as a believer, except the rare icon smasher who believes in belief and who, strangely enough, believes himself or herself to be the only unbeliever. This healthy, wide ranging, popular, indestructible agnosticism may be the source of much confusion because, here again, the reactions they trigger are indistinguishable from those created by the As', Bs', Cs', and Ds' acts of destruction and regeneration. It is so easy to be shocked. Everyone has a capacity for "shockability" that can certainly be applied to different causes, but not in any case emptied or even diminished.

Take the now famous icon of Pope John-Paul II struck to the ground by a meteorite, Maurizio Cattelan's 1999 sculpture *La Nona Ora*: does it demonstrate a healthy irreverence for authority? Is it a typical case of a cheap provocation aimed at blasé Londoners, who expect to be mildly shocked when they go to an art show, but don't really give a damn for the killing of such a boring image as that of the Pope? Is it, on the contrary, a scandalous attempt to wreck the belief of Polish museum visitors when the piece is shown in Warsaw? Or is it, as Christian Boltanski claims, a deeply respectful image showing that, in Catholicism, the Pope is requested to suffer the same breaking, the same ultimate destruction as Christ himself?[65] How is it possible to test this range of interpretations?[66]

A Welcome Cacophony

Through sound as well as image, I want to restore this sense of ambiguity: who is screaming against destruction and why? Are these the lamentations of the eternal philistines shocked to be forced out of their boring and narrow circle of habits? Hear, hear! are these the wailings of humble worshippers deprived of their only source of virtue and attachment—the sacred relics, the precious fetishes, the fragile factishes—that used to keep them alive, and which are now broken by some blind and arrogant reformer?[67] Hear, hear! the weeping sound made by the As realizing that they will never attain the gentle violence of the prophetic Bs, and that they have simply emptied the world and made it even more terrifying. Hear again, behind the cacophonic laments, the sardonic laugh of the blasphemous Es, so healthy, so happy to deploy their juvenile charivari. And behind it all, what is it, this other sound? Hear, hear! the prophetic trumpet waking us out of our deadly attachment to resuscitate a new sense of the beauty, truth, and sanctity of images. But who makes this horrible raucous noise? Hear, hear! what a racket, the blaring sound of the provocateurs, looking for new prey.

Yes, it is a pandemonium: our daily world.

Beyond the Image Wars: Cascades of Images

How can we be sure that this essay is not another iconoclastic display: that I am not asking the reader to descend one more spiral in the inferno of debunking and criticism? How can I be sure that I am not adding another layer of irony—piling disbelief upon disbelief—continuing the task of disenchantment with even more disenchantment? Agreement in these answers should not be the goal, since I am after *iconoclashes*, not certainty. And yet I claim to be able to go beyond the image wars. Always a bold claim this little preposition: beyond. How can I be faithful to it?

By writing about images, objects, statues, signs, and documents in a way that demonstrates the connections they have with other images, objects, statues, signs, and documents, I am trying to claim the Bs against the As, Cs, Ds, and even the Es. Images do count: they are not mere tokens—not because they are prototypes of something away, above, beneath—they count because they allow one to move to another image, exactly as frail and modest as the former one, but different.[68]

Thus, the crucial distinction I wish to draw in this essay is not between a world of image and a world of no-image—as the image warriors would have us believe—but between the interrupted flow of pictures and a cascade of them. By directing the attention of readers to those cascades, I do not expect peace—the history of the image is too loaded for that—but I am gently nudging the reader to look for other properties of the image, properties that religious wars have completely hidden in the dust, blown up by their many fires and furies.

The Opacity of Religious Icons

Take for instance the small and humble *Pieta* in the Museum of Moulins in France (see fig. 11 on page 120): Protestant or later revolutionary fanatics (or maybe vandals), decapitated the Virgin and broke the limbs of the dead Christ, although the scriptures say that "Not a bone of him shall be broken" (John 19:37). A tiny, intact angel, invisible in the picture, holds in sorrow the falling head of the Savior. An iconoclastic gesture, to be sure. But wait! What is a dead Christ if not another broken icon, the perfect image of God, desecrated, crucified, pierced, and ready to be entombed? So the iconoclastic gesture has struck an image that had already been broken.[69] What does it mean to crucify a crucified icon?

Are we not confronted here with a good *iconoclash*? The idol smasher has been redundant since he (for rather obscure reasons, I keep maintaining the masculine for that sort of deed) has smashed a pre-broken icon. But there is a difference between the two gestures: the first one was a deep and old meditation on the weakness of all icons, the second has only added a sort of simple-minded will to get rid of all idols, as if there were idols, and idol worshippers. The image warriors always make the same mistake: they naively believe in naive belief. Has not the idol-breaker only demonstrated his naiveté by imagining that the first was an idol-worshipper, whereas he or she must have been a pretty good icon-breaker? In this tradition, image is always that of a breaching to render the object unfit for normal consumption.[70]

As Louis Marin has argued in a beautiful book, the same is true of Christian religious paintings that do not try to show anything but, on the contrary, obscure the vision.[71] Thousands of little inventions force the viewer, the worshipper, into not seeing what is presented in front of him or her, not, as the defenders of icons often say, by redirecting the attention away from the image to the prototype. There is no prototype to be looked at—this would be Platonism run mad—but only the redirecting of attention to another image.

The Emmaus pilgrims see nothing in their fellow traveler as painted by Caravaggio, but the breaking of the bread reveals what they should have seen, what the viewer can only see by the very dim light the painter has added to the bread, yet it is nothing but a painting. Redirecting attention is always the job those pictures try to do, thus forcing the faithful to move from one image to the next: "He is not here. See the place where they laid him" (Mark 16: 6).

How wrong headed were the image wars: there is not one of those pictures not already broken in the middle. Every icon repeats: *Noli me tangere*, and they are accused by their enemies of attracting too much attention! Are we really going to spend another century naively re-destroying and deconstructing images that are so intelligently and subtly destroyed already?

Isolated, a Scientific Image Has no Referent

The cascade of images is even more striking when one looks at the series assembled under the label of science.[72] An isolated scientific image is meaningless: it proves nothing, says nothing, shows nothing; it has no referent. Why? Because a scientific image, even more than a Christian religious one, is a set of instructions to reach another one down the line.[73] A table of figures will lead to a grid, which will lead to a photograph, which will lead to a diagram, which will lead to a paragraph, which will lead to a statement. The whole series has meaning, but none of its elements has any sense.

In the beautiful examples on astronomy, shown by Galison, you cannot stop at any place in the series if you want to "grasp" the phenomenon they picture, but if you go up or down the whole series, then objectivity, visibility, veridicality will ensue.[74] The same is true of the molecular biology example offered by Rheinberger: in radio labeling, there is nothing to see at any stage, and yet, there is no other way to see genes.[75] Invisibility in science is even more striking than in religion: hence nothing is more absurd than the opposition between the visible world of science and the "invisible" world of religion.[76] Neither can be grasped except by images broken in such a way that they lead to yet other images.

If you wanted to abandon the image and turn your eyes instead to the prototype that they are supposed to figure out, you would see infinitely less.[77] You would be blinded by such an act. Ask a physicist to turn her eyes away from the inscriptions produced by her detectors and she won't detect a thing: she will begin to have an inkling only if she assembles even more inscriptions, even more instrumental results, even more equations.[78] Only down inside the closed walls of her ivory tower does she gain some access to the world "out there."

This paradox of scientific images is again entirely lost by the image

warriors who violently ask us to choose between the visible and the invisible, the image and the prototype, the real world out there and the
made-up, artificial world in here. They cannot understand that the more
arti-factual the inscription, the better its ability to connect, to ally with
others, to generate even better objectivity.

Thus, to request the idol-breakers to smash the many mediators of science, in order to reach the real world out there, better and faster, would
be a call for barbarism, not for enlightenment. Do we really have to spend
another century alternating violently between constructivism and realism, between artificiality and authenticity? Science deserves better than
naive worship and naive contempt. Its regime of invisibility is as uplifting as that of religion and art. The subtlety of its traces requires a new
form of care and attention, a form of spirituality.

Art Is Not To Be Redeemed

Connecting images to images—playing with series of them, repeating
them, reproducing them, distorting them slightly—has been common
practice in art even before the infamous age of mechanical reproduction.
Intertextuality is one of the ways in which the cascading of images is dis-
cernible in the artistic domain: the thick entangled connection that each
image has with all the others that have been produced, the complex relation of kidnapping, allusion, destruction, distance, quotation, parody,
and struggle.[79] Even the simplest connection is so important for a definition of an avant-garde that, once a type of image has been devised, it is
no longer possible for others to produce it in the same fashion.

But there is a more direct relation: in many ways, through the question of mimetic representation, western arts have been obsessed by the
shadows cast by scientific and religious pictures: how can one escape
from the obligation of once again presenting the credos of the faithful? How can one escape from the tyranny of "simply objective," and
"purely representative" quasi-scientific illustrations? Freeing one's gaze
from this dual obligation accounts for a great deal of the inventions of
what is called modern art. And of course "reactionary" critics never tire

of asking for a return to "real presence" to "accurate representation" to "mimesis," and the worship of beauty as if it were possible to turn back the clock.[80]

So here is another paradox, another *iconoclash*: what is it that contemporary art has so forcefully tried to escape? To what target was so much iconoclasm directed, so much asceticism, so much violent and sometimes frenetic energy? Has it been directed at religious icons and their obsession for real presence? They have never been about presenting something other than absence. Has it been directed at scientific imagery? No isolated scientific image has any mimetic power: there is nothing less representational, less figurative, than the pictures produced by science, which are nonetheless said to give us the best grasp of the visible world.[81]

Here, again, we have another case of image wars directing our attention to a completely false target. Many artists have tried to avoid the heavy load of presence and mimesis by avoiding religion and science, which have striven intensely to shun presence, transparency, and mimesis.

How long are we going to judge an image, installation, or object by those other images, installations, and objects they aim at fighting, replacing, destroying, ridiculing, bracketing, or parodying? Is it so essential to art that a long retinue of slaves and victims accompany every piece? Is the distortion of an already existing image really the only option?

Fortunately, there exist many other forms of art, types of installations—devices of all sorts—that do not in any way rely on this negative connection between image and distortion: not because they rely on mimesis, which would restrict the gaze to the most boring type of visual custom, but because what they like most is the transformation of images, the chain of modifications that completely modify the scopic regimes of the classic frozen image, extracted from the flow.[82]

This difference between iconoclast distortion, which always relies on the power of what is destroyed, and a productive cascade of re-representation might explain why Peter Weibel's definition of art, for instance, does not intersect at all with that of someone like Adam Lowe: another *iconoclash* and a very fecund one, visually.

After 9/11

As Christin, Colas, Gamboni, Assmann and many others have shown, there has always been a direct connection between the status of image and politics. Destroying images has always been a carefully planned, elitist, and governed action. There is nothing less popular, spontaneous, and undirected than idol wrecking. Although the word "representation" appears even more vividly in the public sphere than in science, religion, and art, I have not treated iconoclasm in politics as a separate domain.

There is a simple reason for that: in order to rejuvenate the definition of political mediators, it is first essential to go beyond the image wars. Politics is everywhere, yet iconoclasm has become much too cheap when applied to the political sphere. Nowhere more than in politics can the absurd but strident request, "Is it manipulated or is it real?" be heard. It is as if, again, the work of the hands, the careful manipulation, and the human-made mediation, had to be put in one column, and truth, exactitude, mimesis, and faithful representation into another. As if everything that was added to the credit in one column had to be deducted from the other. What strange accounting, which would make politics as well as religion, science, and art, utterly impossible: another case of an impossible application of the second commandment.

The worship of image destruction, the cult of iconoclasm as the ultimate intellectual virtue, the critical mind, the taste for nihilism—all of that may have changed abruptly due to the terrifying event, strangely coded by the figure 9/11, which happens to be the emergency telephone number in the United States. Since September 11, 2001, a state of emergency has been proclaimed on how we deal with images of all sorts—in religion, politics, science, art, and criticism—and a frantic search for the roots of fanaticism has begun.

Nihilism—understood here as the denial of mediators; as the forgetting of the hand at work in the waking of transcendent objects; as the modernist cut between what one does and what one thinks one is doing—could appear as a virtue, a robust quality, a formidable source of innovation and strength, as long as we could apply it to others for real and to us only symbolically. But now, for the first time, it is the U.S., it is us, the Westerners, the courageous idol-breakers, the freedom-fighters who are threatened by annihilation and fanaticism.

In the same way as Hollywood script writers are suddenly finding unbearable the special effects of the horror-movies, which they had concocted because their realities are too vivid and were only bearable when they could not happen, we might find the constant talk of destruction, debunking, critique, exposure, and denunciation, not so funny after all: not so productive, not so protective.

We knew (I knew!) we had never been modern, but now we are even less so—fragile, frail, threatened—that is, back to normal, back to the anxious and careful stage in which the "others" used to live before being "liberated" from their "absurd beliefs" by our courageous and ambitious modernization. Suddenly, we seem to cling with a new intensity to our idols, to our fetishes, to our "factishes," to the extraordinarily fragile ways in which our hand can produce objects, and over which we have no command. We look at our institutions, our public spheres, our scientific objectivity, even our religious ways, everything we loved to hate before, with a somewhat renewed sympathy—with less cynicism, suddenly, and less irony—a worshipping of images, a craving for carefully crafted mediators: what the Byzantine called "economy;" what used to simply be called civilization.

Ah, by the way, how should Moses have written the second commandment had he not misinterpreted it? It is a bit early to know, we first need to hear and see your reactions, but my bet is that a safer reading would be: "Thou shall not freeze-frame any graven image!"

CHAPTER THREE

"Thou Shall Not Freeze-Frame," Or How Not to Misunderstand the Science and Religion Debate

There exists a difficulty that is specific to religion, not as a belief, but as a mode of speech: whenever you write about it, you risk losing the phenomenon to which you wanted to draw attention, and this loss is even greater with the so-called religions of the Book. This is why, so as to try to extirpate myself out of this difficulty, I have chosen to write the following essay as if it were delivered as a lecture, or even as a sermon. Although I draw attention to the religious *speech-act*, it is through the medium of a *written* text, addressed to a now imaginary audience—among whom the reader should accept a place—that I will compare the conditions of felicity of religious speech-acts with those of another type of speech-act, that of science. I have not found a less awkward way to elicit this most important of contrast. And now to begin.

I have no authority whatsoever to talk to you about religion and experience, since I am neither a preacher, nor a theologian, nor a philosopher of religion, nor even an especially pious person. Fortunately, religion might not be about authority and strength but exploration, hesitation, and weakness. If so, then I should begin by putting myself in a position of most extreme weakness. William James, at the end of his masterpiece *Varieties of Religious Experience*, says his form of pragmatism possesses a crass label, that of pluralism.[1] I should state at the beginning of this talk that my label (should I say my stigma?) is even more crass: I have been raised a Catholic, and worse I cannot even speak to my children of what I am doing at church on Sunday. It is from this very impossibility of speaking about a religion to my friends and to my kin that matters to me, that I want to consider here. I want to begin this lecture by this hesitation, this weakness, this stuttering: by this speech impairment. Religion, in my tradition, in my corner of the world, has become impossible to enunciate.[2]

I do not think I could be allowed to start only from such a weakened and negative position. I also have a slightly firmer ground that gives me some encouragement in addressing this most difficult topic. If I have dared answering the invitation to speak, it is also because I have been working for many years on offering interpretations of scientific practice other than common ones. It is clear that, in an argument on science and religion, any change in the way science is considered—however slight, however disputed—will have some consequences on the many ways to talk about religion. Truth production in science, religion, law, politics, technology, economics, etc. is what I have been studying, over the years, in my program to advance toward an anthropology of the modern (or rather non-modern) world. Systematic comparisons of what I call "regimes of enunciation" or "modes of existence" are what I am after, and if there is any technical argument in what follows, it is from this rather idiosyncratic comparative anthropology that they will come from. In a sort of weak analogy with speech-act theory, I've devoted myself to mapping out the "conditions of felicity" of the various activities that in our cultures are able to elicit truth.

I have to note at the beginning that I am not trying to produce a critique of religion. That truth is in question in science and religion is not, for me, in question. Contrary to what some of you, who might know my work on science (most probably by hearsay), might be led to believe, I am interested mainly in the practical conditions of truth-telling and not in debunking religion, after having disputed the claims of science (so it is said). If it were already necessary to take science seriously, without giving it some sort of "social explanation," such a stand is even more necessary for religion: debunkers and iconoclasts simply would miss the point. Rather, my problem concerns how to become attuned to the right conditions of felicity of those different types of "truth-generators."

I don't think it is possible to speak of religion without making clear the form of speech that is adjusted to its type of "predication." Religion, at least in the tradition from which I am writing—namely the Christian one—is a way of preaching, of predicating, of enunciating truth in a certain way; this is why I have to mimic, in writing, the situation of an oration given from the pulpit. It is literally, technically, theologically, a form of news, of "good news": what in Greek was called *evangelios*; what has been translated in English by "gospel." Thus, I am not going to speak of religion in general, as if there existed some universal domain, topic, or problem called "religion," which could allow one to compare divinities, rituals, and beliefs, from Papua New Guinea to Mecca, from Easter Island to Vatican City. The faithful have only one religion, just as a child has only one mother. There is no point of view from which one could compare different religions and still be talking in a religious fashion. As you see, my purpose is not to talk *about* religion, but to talk to you *religiously*, at least religiously enough so that we can begin to analyze the conditions of felicity of such a speech-act, by demonstrating it *in vivo*, tonight, to this virtual audience, what sort of truth condition it requests. Since the topic of this series implies "experience," experience is what I want to generate.

Talking of Religion, Talking from Religion

What I am going to argue is that religion—again in the tradition which is mine—does not speak *of* things, but *from* things—entities, agencies, situations, substances, relations, experiences, whatever is the word— which are highly sensitive to the *ways* in which they are talked about. They are, so to speak, *manners of speech*: John would say Word, *Logos* or *Verbum*. Either they transport the spirit from which they talk, and they can be said to be truthful, faithful, proven, experienced, self-verifiable, or they don't reproduce, don't perform, don't transport what they talk from and immediately, without any inertia, they begin to lie, to fall apart, to stop having any reference, any ground. Either they elicit the spirit they utter and they are true, or they don't and they are worse than false: they are simply irrelevant, parasitical.

There is nothing initially extravagant, spiritual, or mysterious in describing religious talk in this way.[3] We are used to other, perfectly mundane, forms of speech that are evaluated not by their correspondence with any state of affairs, but by the quality of the interaction they generate in the way they are uttered. This experience—and experience is what we wish to share—is common in the domain of "love-talk" and, more broadly, personal relations. "Do you love me?" is not assessed by the originality of the sentence—none are more banal, trivial, boring, rehashed—but by the *transformation* it generates in the listener, as well as in the speaker. In-formation talk is one thing; trans-formation talk is another. When they are uttered, something happens: a slight displacement in the normal pace of things, a tiny shift in the passage of time. You have to decide, to get involved, perhaps commit yourselves irreversibly. We are not only undergoing one experience among others, but a change in the pulse and tempo of experience: *kairos* is the word the Greeks would have used to designate this new sense of urgency.

Before going back to religious talk—in order to displace our usual framework—I wish to extract two features from the experience we all have (I hope) in uttering or listening to sentences that transmit love.

The first frame concerns such sentences judged, not by their *content*—their number of bytes—but by their performative abilities. These are mainly evaluated only by this question: do they produce the thing they talk about, namely *lovers*? I am not so much interested here in love as *eros*, which often requires little talk, but in love as *agapè* to use the traditional distinction. In love's injunction, attention is redirected not to the content of the message, but to the container itself, the person-making. One does not attempt to decrypt the sentence as if it transported a message, but as if it transformed the messengers themselves. Yet, it would be wrong to say that they have no truth value simply because they possess no informational content. On the contrary, although one could not tick Ps and Qs to calculate the truth table of such statements, it is a very important matter—one to which we devote many nights and days—to decide whether they are truthful, faithful, deceitful, superficial, or simply obscure and vague. This is even more the case because such injunctions are in no way limited to the medium of speech: smiles, sighs, silences, hugs, gestures, gaze, postures, everything can relay the argument—yes it is an argument, and a tightly knit one at that. Indeed it is an odd argument, which is largely judged by the *tone* with which it is uttered, its tonality. Love is made of syllogisms whose premises are persons. Are we not ready to give anything and everything to be able to detect truth from falsity in this strange talk, which transports persons and not information? If there is one involvement in truth detection, in trust-building, that everyone shares, it is certainly this ability to detect right from wrong love talk. Thus one of the conditions of felicity we can readily recognize is the existence of different forms of speech—and again it is not just language—that are able to transfer persons, not information, either because they in part produce personhood, or because new states, "new beginnings"—as William James would say—are generated in the persons thus addressed.

The second feature I wish to retain from the specific—and totally banal—performance of love talk is that it seems able to shift the way space is inhabited and time flows. Here again the experience is so widespread that we might overlook its decisive originality. Although it is so common, it is not that often described, except in a few movies by Ingmar Bergman and a few odd novels, because "eros"—Hollywood eros—typically occupies the stage so noisily that the subtle dynamic of *agapè* is rarely noticed. We can share enough of the same experience to capitalize on it later for my analysis: what happens to you, would you say, when you are addressed by love-talk? Very simply put: you were *far away*, and now you are *closer*; lovers seem to have a treasure of private lore to account for the subtle reasons of this shift from distance to proximity. This radical change concerns not only space but also time: you just had the feeling of inflexible and fateful destiny—as if a flow from the past to the ever diminishing present were taking you straight to inertia, boredom, maybe death—and suddenly, a word, an attitude, a query, a posture, a *je ne sais quoi*, and time flows again, as if it were starting from the present, with the capacity to open the future and reinterpret the past: possibility arises; fate is overcome; you breathe; you feel enabled; you hope; you move. In the same way as the word "close" captured the different ways space is now inhabited, it is the word "present" that now seems the best way to capture what happens to you: you are present *again* and *anew* to one another. Of course you might become *absent* and *far away* again in a moment—this is why your heart beats so fast, why you are at once so thrilled and so anxious: a word badly uttered, a clumsy gesture, a wrong move and, instantly, the terrible feeling of estrangement and distance, this despondency that comes from the fateful passage of time—and all of that boredom falls over you again, intolerable and deadly. You suddenly don't understand what you are doing with one another; it is unbearable, simply unbearable.

Have I not sketched a very common experience, the one acquired in the love crisis, on both sides of the infinitely small difference between what is close and present and what is far and absent? This difference is marked by a nuance: sharp as a knife, both subtle and sturdy; a difference between talking rightly and talking wrongly about what makes us alive to the presence of another?

If we combine the two features of lovers' speech I have just outlined, we may convince ourselves that a form of speech exists that is concerned by the transformation of messengers instead of the transport of information. A form that is so sensitive to the tone in which it is uttered that it can abruptly shift, through a decisive crisis, from distance to proximity—and back to estrangement—from absence to distance and back again. Of this form of talk, I will say that it re-presents in one of the many literal meanings of the word: it presents anew what it is to be present at what one says. Further, this form of talk is at once completely common, extremely complex, and not that frequently described in detail.

How is Attention Redirected?

Such is the atmosphere from which I wish to benefit in order to start my predication again: since to talk, nay to preach, religion is what I am attempting to do, so as to obtain enough common experience that it can be analyzed afterwards. I use the template of lovers' speech in order that we may rehabituate ourselves to a form of religious talk that has been lost, unable to represent itself again—to repeat itself—because of the shift from religion to belief. I will address this more thoroughly later. We now know that the competence we are looking for is common, that it is subtle, that it has not been thoroughly described, that it easily appears and disappears, that it tells the truth and then lies. The conditions of felicity of my own talk are thus clearly outlined: I will fail if I cannot produce, perform, or educe what it is about. Either I am able to re-present it to you again—that is, present it in its renewed and olden presence—and I speak in truth, or I don't, and although I might have pronounced the same words, it is in vain that I speak: I have lied to you; I am nothing but an empty drum that beats in the void.

Three words are important, then, to respect my risky contract with you: "close," "present," and "transformation." To give myself some chance to succeed in re-enacting the right way to say religion things— according to the tradition of the Word in which I have been raised—I need to direct your attention away from topics and domains thought to pertain to religion, but which might render you indifferent or hostile to my way of talking. We have to resist two temptations in order for my argument to stand a chance of representing anything, and thus to be truthful. The first temptation is the abandonment of the transformation necessary for this speech-act to function; the second would direct our attention to the *far away* instead of to the close and present.

To put it simply, but I hope not too provocatively, if, when hearing about religion, you direct your attention to the *far away*—the above, the supernatural, the infinite, the distant, the transcendent, the mysterious, the misty, the sublime, the eternal—chances are that you have not even begun to be sensitive to that in which religious talk tries to involve you. Remember, I am using the template of lovers' speech, to speak of different sentences with the same spirit, the same regime of enunciation. In the same way as those love sentences should transform the listeners, either into being close and present or else void, the ways of talking religion should bring the listener, and also the speaker, to the same closeness and to the same renewed sense of presence: otherwise they are worse than meaningless. If you are attracted to the distant, by religious matters, to the far away, the mysteriously encrypted, then you are gone:

literally you are not with me; you remain absent minded. You make a lie of what I am giving you a chance to hear again. Do you understand what I am saying and the way I am saying it: the Word tradition I am setting into motion again?

The first attempt at redirecting your attention is to make you aware of the pitfalls of double-click communication. If you use such a bench mark to evaluate the quality of religious talk, it will become exactly as meaningless, empty, boring, and repetitive as misaddressed lovers' speech, and for the same reason, since it carries no messages, but transports and transforms the messengers themselves, or it fails to do so. And yet, this is exactly the yardstick of double-click communication: it wants us to believe that it is feasible to transport, without any deformation whatsoever, some accurate information about states of affairs that are not present to us. In most ordinary cases, what people have in mind when they ask "is this true," or "does this correspond to a state of affairs," is just such a double-click gesture, allowing immediate access to information: tough luck, because this is also what undermines ways of talking that are dearest to our heart. On the contrary, to disappoint the drive towards double-click—to divert it, to break it, to subvert it, to render it impossible—is just what religious talk is after. It wants to make sure that even the most absent-minded, most distant gazers are brought back to attention, so that they don't waste their time ignoring the call to conversion: to disappoint, first, to disappoint. "What has this generation done in requesting a sign? No sign will be given to them!"

Transport of information, without deformation, is not one of religious talk's conditions of felicity. When the Virgin hears the angel Gabriel's salutation, she is so utterly transformed that she becomes pregnant with the Savior, rendered, through her agency, present again to the world. Surely this is not a case of double-click communication! Yet, asking who Mary was, checking whether or not she was really a Virgin, imagining a pathway to impregnate her with spermatic rays, deciding whether Gabriel is male or female, are double-click questions. They want you to abandon the present time and to direct your attention away from the meaning of the venerable story. These questions are not impious, nor even irrational: they are simply a category mistake. They are so irrelevant that no one has even bothered answering them. Not because they lead to unfathomable mysteries, but because their idiocy makes them generate uninteresting and utterly useless mysteries. They should be broken, interrupted, voided, or ridiculed, and I will show later how this interruption has been systematically attempted in one of the Western Christian iconographic traditions. The only way to understand stories, such as that of the Annunciation, is to repeat them, that is, to utter again a Word that produces in the listener the same effect, namely one that impregnates with the gift of renewed presence. Here and now I am your Gabriel, or else you don't understand a word of what I am saying, and I am a fraud.

Not an easy task—I will fail, I am bound to fail, I speak against all odds—but my point is different because it is a little more analytical: I want you to realize through what sort of category mistake belief in belief is being generated.[4] Either I repeat the first story because I retell it in the same efficient mode in which it was first told, or I hook up a stupid referential question to a message (in terms of information transfer), and I am worse than crassly stupid: I make the venerable story lie because I have distorted it beyond recognition. Paradoxically, by formatting questions in the Procrustean bed of information transfer in order to get at its "exact" meaning, I would deform the story, transmogrify it into an absurd belief, into the sort of belief that weighs religion down and lets it slide towards the refuse heap of past obscurantism. The truth value of such stories depends on us, right now, exactly as the whole history of two lovers depends on their ability to re-enact the injunction to love again, in the moment they are reaching for one another and in the darker moment of their estrangement: if they fail (present tense), it was in vain (past tense) that they have lived so long together.

Note that I did not speak of those sentences as being either irrational or unreasonable, as if religion needed somehow to be protected against an irrelevant extension of rationality. When Ludwig Wittgenstein writes, "I want to say 'they don't treat this as a matter of reasonability.' Anyone who reads the Epistles will find it said: not only that it is not reasonable, but that it is a folly. Not only it is not reasonable, but it doesn't pretend to be,"[5] he seems to deeply misunderstand what sort of folly the Gospel is writing about. Far from not pretending to be reasonable, it simply applies the same common reasoning to a different kind of situation: it does not try to reach distant states of affairs, but brings the interlocutors closer to what they say of one another. To suppose that, in addition to the rational knowledge of what is graspable, there also exists some sort of non-reasonable and respectable belief of things too far away to be graspable, seems to me a very condescending form of tolerance. I'd rather like to say that rationality is never in excess, that science knows no boundary, and that there is absolutely nothing mysterious, or even unreasonable in religious talk—except the artificial mysteries generated, as I just said, by asking the wrong questions—in the wrong mode; in the wrong key—to perfectly reasonable person-making arguments. To seize something by talk—or to be seized by someone else's talk—might be different, but the same basic mental, moral, psychological, and cognitive equipment is necessary for both.

More precisely, we should differentiate two forms of mysteries: one which refers to the common, complex, and subtle ways in which one has to pronounce love talk for it to be efficacious—and it is indeed a mystery of ability, a knack, like good tennis, good poetry, good philosophy, or maybe a sort of folly—and another mystery, totally artificial, which is caused by the undue short-circuit of two different regimes of enunciation colliding with one another. The confusion between the two mysteries is what makes the voice of people quiver when they talk of religion, either because they wish to have no mystery at all—good, there is none anyway—or because they believe they are looking at some encrypted message that they have to decode through the use of some esoteric grid only initiates know how to use. But there is nothing hidden, nothing encrypted, nothing esoteric, nothing odd in religious talk: it is simply difficult to enact; it is simply a little bit subtle; it needs exercise; it requires great care; it might save those who utter it. To confuse talk that transforms messengers with talk that transports messages—cryptic or not—is not proof of rationality: it is simply idiocy doubled by impiety. It is as idiotic as if a lover, asked to repeat whether she loves her partner or not, simply pushes the play-button of a tape recorder in order to prove that, five years ago, she had indeed said "I love you darling." It might prove something, but certainly not that she has renewed her pledge to love presently—it is a valid proof, to be sure: a proof that she is absent minded and probably a lunatic.

That is all I have to say about double-click communication. The two other features—closeness and presence—are much more important for our purposes because they will lead us to the third term of our argument, namely science. It is amazing that most speakers, when they want to show generosity toward religion, have to couch it in terms of its necessary irrationality. I sort of prefer those who, like Pascal Boyer, frankly want to explain—to explain away—religion altogether, by highlighting the brain loci and the survival value of some of its most barbarous oddities.[6] I always feel more at home with purely naturalistic accounts than with this sort of hypocritical tolerance, which ghettoizes religion into a form of nonsense, specialized in transcendence and feel-good inner sentiment. Alfred North Whitehead put an end, in my view, to those who wish religion to "embellish the soul" with pretty furniture.[7] Religion, in the tradition I want to render present again, has nothing to do with subjectivity, nor with transcendence, nor with irrationality. The last thing it needs is tolerance from open-minded and charitable intellectuals, who want to add to the true but dry facts of science the deep and charming supplement of the soul provided by quaint religious feelings.

Here I am afraid I have to disagree with most, if not all, of the former speakers of the science–religion confrontation, because they speak like Camp David diplomats drawing lines on maps of the Israeli and Palestinian territories. They try to settle disputes as if there was one single domain, one single kingdom to share in two, or—following the terrifying similarity with the Holy Land—as if two equally valid claims had to be established side by side: one for the natural, the other for the supernatural. And for some speakers, who, like the most extremist zealots of Jerusalem and Ramallah—the parallel is uncanny—reject the efforts of diplomats and want to claim the entire land for themselves, either by driving the obscurantist religious folks to the other side of the Jordan river, or, conversely, by drowning the naturalists into the Mediterranean sea. I find those disputes—whether there is one or two domains, whether it is hegemonic or parallel, whether polemical or peaceful—equally moot for this reason: They all suppose that science and religion have similar but divergent claims in reaching and settling a territory, either of this world or of some other world. I believe on the contrary that there is no point of contact between the two, no more than, say, nightingales and frogs have to enter into any sort of direct ecological competition.

I am not saying that science and religion are incommensurable—because one grasps the objective visible world of here and there, while the other grasps the invisible, subjective, or transcendent world of beyond—but that even their incommensurability would be a category mistake. The reason is that neither science nor religion fits even a basic picture that would put them face to face, that is, enough in relation to be deemed incommensurable! Neither religion nor science is much interested in the visible: it is science that grasps the far and distant; as to religion, it does not even try to *grasp* anything.

Science and Religion: A Comedy of Errors

My point might appear at first counterintuitive, because I wish to draw simultaneously on what I have learned from science studies about scientific practice and what I hope you have experienced in this lecture's reframing—with the help of love—of religious talk. Religion does not even try, if you have followed me until now, to reach anything beyond, but to represent the presence of that which is called, in a certain technical and ritual idiom, the "Word incarnate," which is to *say* again that it is here—alive—and not dead over there, far away. It does not try to designate something, but to speak from a new state that it generates by its ways of talking, its manner of speech. Religion, in this tradition, does everything to constantly redirect attention, by systematically breaking the will to go away, ignore, be indifferent, blasé, or bored. Conversely, science has nothing to do with the visible, the direct, the immediate, the tangible, or the lived world of common sense: of sturdy "matters of fact."

It is quite the opposite, as I have shown many times: it builds extraordinarily long, complicated, mediated, indirect, and sophisticated paths so as to reach the worlds—like James I insist on the plural—that are invisible because they are too small, too far, too powerful, too big, too odd, too surprising, too counterintuitive, through concatenations of layered instruments, calculations, and models. Only through the laboratory and instrument networks can you obtain those long referential chains that maximize the two contrary features of mobility (or transport) and immutability (or constant) that make up in-formation, which I therefore called "immutable mobiles."

Science in action, science as it is done practically, is even further from double-click communication than religion. Distortion; transformation; recoding; modeling; translating: each of these radical mediations is necessary to produce reliable and accurate information. If science were in-formation *without* transformation, as good common sense would like to have it, we would still be in complete obscurity about states of affairs distant from the here and now. Double-click communication does even less justice to the transformation of information in scientific networks than it does to the strange ability of some speech-acts to transform the interlocutors in religion.

What a comedy of errors! When the debate between science and religion is staged, adjectives are almost exactly reversed: it is science that reaches the invisible world of beyond, that she is spiritual, miraculous, soul-fulfilling, and uplifting;[8] it is religion which should be qualified as being local, objective, visible, mundane, un-miraculous, repetitive, obstinate, and sturdy.

In a fable of the race between the scientific rabbit and the religious tortoise, two things are totally unrealistic: the rabbit *and* the tortoise. Religion does not even attempt to race for knowledge of the beyond, but attempts to break all habits of thought that direct our attention to the far away—to the absent, to the over-world—in order to bring attention back to the incarnate, to the renewed presence of what was previously misunderstood, distorted, and deadly—of what was, what is, what will be— toward those words that carry salvation. Science does not *directly* grasp anything accurately, but slowly gains its accuracy, its validity, its condition of truth by the long, risky, and painful detour through mediations: of experiments, not experience; laboratories, not common sense; theories, not visibility. If it obtains truth, it is at the price of mind-boggling transformations from one media into the next. Thus, even to assemble a stage, in which the deep and serious problem of "the relation between science and religion" could unfold, is already an imposture—not to say a farce—that distorts science and religion beyond all recognition.

The only protagonist who would dream of the silly idea of staging a race between the rabbit and the tortoise, to bring them face to face, so as to decide afterward who dominates whom—or to invent even more bizarre diplomatic settlements between the two characters—the only Barnum for such a circus, is double-click communication. Only he, with this bizarre idea of transportation *without* transformation—in order to reach a faraway state of affairs—could dream of such a confrontation, distorting the careful practice of science as well as the careful repetition of person-forming religious speech. Only he can make both science and religion incomprehensible, first by distorting the mediated and indirect access of science to the invisible world—through the hard labor of scientists—into a direct, plain, and unproblematic grasp of the visible; then, in giving the lie to religion by forcing it to abandon its goal of representing anew what it is about, making all of us gaze, absent-mindedly towards the invisible world of beyond that it has no equipment, nor competence, nor authority, nor ability to reach, and even less to grasp. Yes, what a comedy of errors, a sad comedy, which has made it almost impossible to embrace rationalism, since that would mean ignoring the workings of science even more than the goals of religion.

Two Different Ways of Linking Statements to One Another

These two regimes of invisibility, which have been so distorted by an appeal to the dream of instant and unmediated communication, might be made more demonstrative by appealing to visual documents. My idea, as I hope it is now clear, is to move the reader from one opposition between science and religion, to another one between two types of objectivity. The first, traditional, fight has pitted science—defined as the grasp of the visible, the near, the close, the impersonal, the knowable—against religion, which is supposed to deal with the far, the vague, the mysterious, the personal, the uncertain, and the unknowable.

To this opposition, which is an artifact, I want to substitute another opposition between the long and mediated referential chains of science—that lead to the distant and the absent—and the search for the representation of the close and present in religion. As I have shown elsewhere science is in no way a form of speech-act that tries to bridge the abyss between words and the world (in the singular). That would amount to the *salto mortale*, so ridiculed by James; rather science as it is practiced, attempts to deambulate—James's expression again—from one inscription to the next, by taking each of them in turn for the matter out of which it extracts a form, which here has to be meant very literally, very materially: it is the paper in which you place the "matter" of the stage just preceding.

Since an example is always better to render visible the invisible path that science traces through the pluriverse, let's take the case of Jean R's laboratory in Paris, where attempts are made to gain information on the releasing factors of one single isolated neuron. Obviously, there is no unmediated, direct, non-artificial way to render one neuron visible out of the billions that make up grey matter, so they have to begin with rats. First they are guillotined: the brain is extracted then cut, thanks to a microtome, into very fine slices; each slice is prepared in such a way that it remains alive for a couple of hours. It is put under a powerful microscope and then, on the screen of the television, a micro-syringe and a micro-electrode are delicately inserted into one of the neurons, on which the microscope is able to focus, among the millions that are simultaneously firing. All of this may fail. Focusing on one neuron and bringing the micro-syringe into contact with the same neuron, in order to capture the neurotransmitters while recording the electric activity, is a feat few people are able to achieve. The neural activity is recorded and the chemical products triggered by its activity are gathered through a pipette. The findings are written into an article that synoptically presents the various inscriptions. I don't want to say anything about neural firing—no matter how interesting—but to attract your attention to the movement, the jump from one inscription to the next.

It is clear that without the artificiality of the laboratory, none of this path through inscriptions, in which each plays the role of matter for the next—transforming it—produces a *visible* phenomenon. Reference is not the gesture of an interlocutor pointing with a finger to a cat purring on a mat, but a much riskier affair and a much dirtier business, which connects a published literature outside the lab, to published literature from the lab, through many intermediations, one of them being the rats, those unsung heroes of biology.

The point I want to make is that these referential chains have very interesting contradictory features: they are producing our best source of objectivity and certainty, yet they are artificial, indirect, and multilayered. There is no doubt that the reference is accurate, but this accuracy is not obtained by any two things resembling each other mimetically, but through whole chains of artificial and highly skilled *transformations*. As long as the chain succeeds, the truth-value of the whole reference is calculable. But if you isolate one inscription, extract one image, or freeze-frame the continuous path of transformations, then the quality of the reference immediately deteriorates. Isolated, a scientific image has no truth value, although it might trigger—in the mythical philosophy of science that predominates—a sort of shadow referent, which is taken, by a sort of optical illusion, to be the model of the copy, although it is nothing but the virtual image of an isolated copy!

This proves that matters of fact—those famous matters of fact that are supposed by some philosophers to be the stuff out of which the visible common sense world is made—are actually nothing but a misunderstanding of the artificial but productive process of scientific objectivity: what has been derailed by freeze-framing a referential path. There is nothing primitive, nothing primeval in matters of fact: facts are not the ground of mere perceptions.[9] Thus it is entirely misguided to try to *add* to the objective matters of fact some sort of subjective state of affairs that, in addition, would occupy the mind of the believers.

Although some of what I have said here—much too briefly—might still be controversial, let us bracket it as if it were an undisputed background, thereby it may shed new light on the religious regime of invisibility. In the same way as there is a misunderstanding on the path traced by the deambulation of scientific mediations, there is, I think, a common misunderstanding on the path traced by religious images.[10] The traditional defense of religious icons in Christianity has been to say that the image is not the object of a *latry*, but of a *dulia*, a Greek term to say that a worshipper, at the occasion of the copy—a Virgin, a crucifix, the statue of a saint—turns his or her mind to the prototype, the only original worth adoring. This is, however, a weak defense, which never convinced the Platonist, the Byzantine, the Lutheran, or the Calvinist iconoclasts—not to mention Mullah Omar when he had the Bamihyan Buddhas put to the gun.

In effect, the Christian regime of invisibility is as different from this traditional meek defense as the scientific reference is from glorified matters of fact. What imagery has tried to achieve through countless feats of art is the opposite of turning the spectator's eyes to the model far away: on the contrary, incredible pains have been taken to *break* the habitual gaze of the viewer, so as to attract attention to the *present* state, the only one which can be said to offer salvation. It is as if the painters, carvers, and patrons of works of art had tried to break the interiority of images, so as to render them unfit for normal informative consumption; as if they wanted to begin, to rehearse, to start a rhythm, a movement of conversion that is understood only when the viewer—the pious viewer—takes it upon himself to repeat the same tune in the same rhythm and tempo. This is what I call, with Joseph Koerner, "inner iconoclasm," compared to which the "external" iconoclasm looks always at least naive and moot, not to mention plain silly.[11]

A few examples will be enough. In Fra Angelico's fresco in San Marco, Florence, the painter has multiplied the ways of complicating our direct access to the topic: not only is the tomb empty—first a great disappoint-

ment to the women—but the angel's finger points to an apparition of
the resurrected Christ, which is not directly visible to the women because it is behind them. What can be more disappointing and surprising
than the angel's utterances: "He is no longer here, he has risen." Everything in this fresco points to the emptiness of the mundane grasp. It is
not, however, *about* emptiness, as if one's attention were directed toward
nothingness; it is, on the contrary, slowly bringing us back to the presence of presence: for that we should not look at the painting, at what
the painting suggests, but at what is now present for us. How could one
evangelist, and then a painter like "brother Angelic," better render vividly
the redirection of attention: "You look in the wrong place . . . you have
misunderstood the Scriptures." In case we are dumb enough to miss the
message, a monk placed on the left—the representative of the occupant
of the cell—will serve as a *legend* (in the etymological sense of the word)
of the whole story, that is, he will show us how we should see: what does
he see? Nothing at all, there is nothing to see *there*, but you should look
here, through the inward eye of piety to what this fresco is supposed to
mean: elsewhere, but not in a tomb, not among the dead but among the
living.

Ever more bizarre is the case, studied by Louis Marin, of the "Annunciation," by Piero della Francesca in Perugia.[12] If you reconstruct the picture in virtual reality—and Piero was such a master at this first mathe-
matization of the visual field that it can be done very accurately with a
computer—you realize that the angel actually remains invisible to the
Virgin! He—or she—remains hidden by the pillar. This cannot be an
oversight. Piero has used the powerful tool of perspective to recode his
interpretation of what an invisible angel is, so as to render impossible
the banal, usual, and trivial *view* that this is a normal messenger meet-
ing the Virgin in the normal space of daily interactions. Again, the idea
is to avoid, as much as possible, the normal transport of messages, even
when using the fabulous new space of visibility and calculation invented
by Quattrocento painters and scientists: the same space that will be put
to use so powerfully by science, to multiply those immutable mobiles I
discussed earlier. The aim is not to add an invisible world to the visible

one, but to distort, to opacify the visible world enough that one is not led to misunderstand the Scriptures, but to re-enact them truthfully.

To paint the disappointment of the visible without simply painting another world of the invisible—which would be a contradiction in terms—no painter is more astute than Caravaggio. In his famous rendering of the Emmaus pilgrims, who do not understand at first that they have been travelling with the resurrected Savior and recognize him only when he breaks the bread at the inn table, Caravaggio re-produces this very invisibility through a tiny spot of light that redirects the attention of the pilgrims when they suddenly realize what they must see. Of course, the whole idea of painting such an encounter, without adding any supernatural event, is to redirect the attention of the viewer of the *painting*, who suddenly realizes that he or she will never see more than those tiny breaks—these paint strokes—and that the reality they have to turn to is not absent in death—as the pilgrims were discussing along the way coming to the inn—but present now in its full *and veiled* presence. The idea is not to turn our gaze away from this world to another world beyond, but to grasp at last, at the occasion of this painting, this miracle of understanding: what is in question in the Scriptures is now realized between the painter, the viewers, the patrons, and among you: have you not understood the Scriptures? He has risen, why do you look far away in death, it is here, it is present anew. "This is why our heart burnt so much while he was talking."

Christian iconography, in all its forms, has been obsessed by this question of representing anew its concerns, making sure (visually) that there is no misunderstanding in the messages transmitted, that it is really a messenger that transforms what is in question in the speech-act, and not a mere message-transfer wrongly addressed.

In the venerable and somewhat naive theme of the "Mass of St. Gregory"—banned after the Counter Reformation—the argument seems much cruder than in Caravaggio, but it is deployed with the same subtle intensity. Pope Gregory is suddenly supposed to have seen, while celebrating the mass, the host and the wine replaced in three dimensions by the real body of the suffering Christ, with all the associated instruments of the Passion. Real presence is here represented yet again, then painted in two dimensions by the artist, to commemorate this act of re-understanding by the Pope *realizing*, in the full sense of the verb, what the venerable ritual meant.

This rather gory imaging will become repulsive to many after the Reformation, but the point I want to make is that each of those pictures, no matter how sophisticated or naive, canonical or apocryphal, always sends a *double* injunction: the first one concerns the *theme* they illustrate, and most of those images, like the lovers' speech with which I began, are often boringly repetitive; but then they send a *second* injunction that traverses the boring repetition of the theme and forces us to remember what it is to understand the presence that the message is carrying. This second injunction is equivalent to the tone, to the tonality of which we have been made aware in lovers' speech: it is not what you say that is original, but the movement that renews the presence through the old sayings.

Lovers, as well as religious painters and their patrons, carefully have to make the usual way of speaking vibrate in a certain way if they want to ensure that the absent-minded interlocutors are not led far away in space and time. This is exactly what happens to poor Saint Gregory: during the repetition of the ritual, he is suddenly struck by the very speech-act of transforming the host into the body of Christ, by the realization of the words under the form of a suffering Christ. It would be a mistake to think this is a naive image that only backward papists could take seriously: quite the opposite, it is a very sophisticated rendering of what it is to become aware again of the real presence of Christ in the mass. But to arrive at this point, you have to listen to the *two* injunctions at once. This is not the painting of a miracle, although it is also that, rather this painting states what it is to understand the word "miracle" literally, not in the habitual, blasé sense of the word, but here meaning not the opposite of spiritual, but ordinary, absent-minded, indifferent.

Even an artist so brilliant as Philippe de Champaigne, in the middle of the seventeenth century, was still making sure that no viewer could ignore that repeating of the face of Christ, literally printing it on a veil;

the effect should *not* be confused with a mere photocopy. This extraordinary meditation on what it is to hide and to repeat is revealed by the presence of three different linens: the cloth out of which the canvas is made, doubled by the cloth of what is called a Veronica, tripled by another veil — a curtain — this one in *trompe l'œil*, which could dissimulate the relic with a simple gesture of the hand, if one were silly enough to misunderstand its meaning. How magnificent to call what is exactly a *false* picture thrice veiled, *vera icona* ("true image" in Latin): it is so impossible to take it as a photograph that, by a miracle of reproduction, a *positive* and not a negative of Christ's face is presented to the viewer: those artists, printers, and engravers knew everything about positive and negative, so again, as in the case of Piero, this cannot be an oversight. Of course this is a "false positive" — if I may use this metaphor — since the *vera icona*, the true picture, is precisely not a reproduction in the referential meaning of the word, but a reproduction in the re-presentational sense of the word: "Beware! Beware! To see the face of Christ is not to look for an original — for a true referential copy that would transport you back to the past, back to Jerusalem — but a mere surface of cracking pigment, a millimeter thick, which begins to indicate how you now, in this Port Royal institution, should look at your Savior." Although this face seems to look back at us so plainly, it is even more hidden and veiled than God's refusal to reveal his own to Moses. To show *and to hide* is what true reproduction does, on the condition that it is false according to the standard of photocopies, printing and double-click communication. What is hidden is not a secret message beneath the apparent one, esoteric information disguised in the banal, but a tone, an injunction for you the viewer to redirect your attention, away from the dead and back to the living.

This is why there is always some uncertainty when a Christian image has been destroyed or mutilated. This Pieta (fig. 11) was broken, likely by some fanatic — we do not know if it was during the Reformation or during the Revolution, France has not lacked in such episodes — but whoever he was, he certainly never realized how ironic it could be to add an *outer* destruction to the *inner* destruction that the statue represented so well: what is a Pieta, if not the image of a heartbroken Virgin holding on

FIGURE 11.
"Pieta," Musée Anne de Beaujeu, Moulins, Allier, 15° siècle Collection Tudot.

her lap the broken corpse of her son, who is the broken image of God his father? How can one destroy an image that is already destroyed? How might you eradicate belief in an image that has already disappointed all beliefs to the point that God himself, the God of beyond and above, lies dead on his mother's lap? Can one go further into the self critique of all images than what theology explicitly says? Who is more naive, the one who sculpted this Pieta of the *kenosis* of God, or the one who believes that there are believers naive enough to grant existence to a mere image, instead of turning their gaze spontaneously to the true original God? Who goes further? Probably the one who says there is *no* original.

How Do Truth-Making Statements Continue to Move?

In conclusion, one way to summarize my point is to say that we have been mistaken in defending the images by their appeal to a prototype to which they simply alluded, although this is, as I showed above, the traditional defense of images. Iconophily has nothing to do with looking at the prototype as a sort of platonic stair-climbing. Rather, iconophily *continues* the process begun by the image, as a prolongation of the flow of images. Saint Gregory continues the text of the Eucharist when he sees the Christ in his real—not symbolic—flesh, and the painter continues the miracle when he paints a representation that reminds us of what it is to understand what this old mysterious text is about. Today, I continue the painter's continuation of the story in reinterpreting the text: if—by using images, arguments, tone, anything at hand—I make you aware again of what it is to understand those images without searching for a prototype, and without distorting them through so many information-transfer vehicles. Iconoclasm or iconolatry are nothing but freeze-framing, interrupting the movement of the image and isolating it out of its flows of renewed images, in order to believe it has a meaning by itself—and since it has none, once isolated, it should be destroyed without pity.

By ignoring the *flowing* character of science and religion, we have turned the question of their relations into an opposition between knowledge and belief, an opposition that we then deem necessary, either to overcome—to politely resolve—or to widen violently. What I have argued in this lecture is very different: *belief is a caricature of religion exactly as knowledge is a caricature of science.* Belief is patterned after a false idea of science, as if it were possible to raise the question "Do you believe in God?" in the same way as "Do you believe in global warming?" except the first question does not possess any of the instruments that would allow the reference to move on, and that the second is leading the interlocutor to a phenomenon even more invisible to the naked eye than God, since to reach it we have to travel through satellite imaging, computer simulation, theories of earth atmospheric instability, or high stratosphere chemistry. Belief is not a quasi-knowledge question *plus* a leap of faith to reach even *further* away; knowledge is not a quasi-belief question that would be answerable by looking directly at things close at hand.

In religious talk, there is indeed a leap of faith, but this is not an acrobatic *salto mortale*, in order to do even better than reference, with more daring and riskier means: it is a somersault yes, but which aims at jumping, dancing towards the present and the close: to redirect attention away from indifference and habituation, to prepare oneself to be seized again by this presence that breaks the usual, habituated passage of time. As to knowledge, it is not a direct grasp of the plain and the visible against all beliefs in authority, but an extraordinarily daring, complex, and intricate confidence in chains of nested transformations of documents that, through many different types of proofs, lead away toward new types of visions, which force us to break away from the intuitions and prejudices of common sense. Belief is simply immaterial for any religious speech-act; knowledge is not an accurate way to characterize scientific activity. We might move forward a bit, if we were calling *faith* the movement that brings us to the close and to the present, and retaining the word belief for this necessary mixture of confidence and diffidence with which we need to assess all the things we cannot see *directly*. The difference between science and religion then would not be found in the different mental competences brought to bear on two different realms — belief applied to vague spiritual matters, and knowledge to directly observable things — but in the *same* broad set of competences applied to *two* chains of mediators going in two *different* directions. The first chain, the scientific one, leads toward what is invisible, because it is simply too far and too counterintuitive to be grasped directly; the second chain, the religious one, also leads to the invisible, but what it reaches is not invisible, not because it would be hidden, encrypted, and distant, but simply because it is difficult to renew.

What I mean is that in the cases of both science and religion — freeze-framing, isolating a mediator out of its chains, out of its series — instantly forbids the meaning to be carried in truth. Truth is not to be found in correspondence — either between the word and the world in the case of science, or between the original and the copy in the case of religion — but in taking up again the task of *continuing* the flow, of elongating the cascade of mediations one step further. My argument is that, in our present economy of images, we might have made a slight misunderstanding on Moses' second commandment, and we might have lacked respect for mediators. God did not ask us not to make images (what else do we have to produce objectivity, to generate piety?) but he told us not to freeze-frame, that is, not to isolate an image out of the flows that only provide them with their real—their constantly re-realized, re-represented—meaning.

I have most probably failed in extending this flow, this cascade of mediators to you in this talk. If so, then I have lied, I have not been speaking religiously, I have not been able to preach, but I have simply talked *about* religion, as if there was a domain of specific beliefs one could relate to by some sort of referential grasp. This then would have been a mistake, just as great as that of the lover who, when asked "do you love me?" answered, "I have already told you so many years ago, why do you ask again?" Why? Because it is no use having told me so in the past, if you cannot tell me again, now, and make me alive to you again, close and present anew. Why would anyone claim to speak religion if it is not in order to save me, to convert me, on the spot?

Notes

Preface

1. *Petite réflexion sur le culte moderne des dieux faitiches* (Paris: Les Empêcheurs de penser en rond, 1996), republished by La Découverte in 2009.
2. "What is Iconoclash? Or Is There a World Beyond the Image Wars?" *Iconoclash: Beyond the Image-Wars in Science, Religion and Art*, ed. Bruno Latour and Peter Weibel, 14–37 (Cambridge, Mass.: MIT Press, 2002).
3. "'Thou Shall Not Freeze-Frame' — or How not to Misunderstand the Science and Religion Debate." *Science, Religion and the Human Experience*, ed. James. D. Proctor, 27–48 (Oxford: Oxford University Press, 2005).

On the Cult of the Factish Gods

1. The Portuguese dictionary *Aurélio* provides the following definitions (N. B.: the Portuguese *feitiço* itself comes from French, through President de Brosses):

   — *feitiço* [from feito + iço] 1. adj. artificial, factício; 2. postiço, falso; 3. Malefício de feticaros; 4. *see* bruxaria [sorcery, witchcraft]; 5. *see* fétiche; 6. encanto, fascinação, fascínio. Proverb: "virar o feitiço contra o feiticeiro";
   — *feitio* [from feito + io] forma, figura, configuração, faição;
   — *fétiche* 1. objeto animado ou inanimado, feito pelo homem ou produzido pela natureza, aqual se atribui poder sobrenatural e se presta culto, ídolo, manipanso.

   Note the marvelous Italian term that gives the same verb, *fatturare*, meaning 1. to falsify, adulterate; 2. to bill; 3. to bewitch!
2. Charles de Brosses, "Du culte des dieux fétiches," in *Corpus des oeuvres de philosophie de langue française*, 2nd edn. (Paris: Fayard, 1988 [1760]), 15. De Brosses's etymology has not been repeated anywhere else. Could there be some contamination between fairy and fetish words?

3. William Pietz offers an excellent summary of de Brosses's invention:

> Fetishism was a radically novel category: it offered an atheological ex-planation of the origin of religion, one that accounted equally well for theistic beliefs and nontheistic superstitions; it identified religious superstition with false causal reasoning about physical nature, making people's relation to material objects rather than to God the key ques-tion for historians of religion and mythology; and it reclassified the entire field of ancient and contemporary religious phenomena . . . In short, the discourse about fetishism displaced the great object of En-lightenment criticism—religion—into a causative problematic suited to its own secular cosmology, whose 'reality principle' was the absolute split between the mechanistic-material realm of physical nature (the blind determinisms of whose events excluded any principle of teleo-logical causality, that is, Providence) and the end-oriented human realm of purposes and desires (whose free intentionality distinguished its events as moral action, properly determined by rational ideals rather than by the material contingency of merely natural being). Fetishism was the definitive mistake of the pre-enlightened mind: it supersti-tiously attributed intentional purpose and desire to material entities of the natural world, while allowing social action to be determined by the (clerically interpreted) wills of contingently personified things, which were, in truth, merely the externalized material sites fixing people's own capricious libidinal imaginings ('fancy' in the language of that day). ("Fetishism and Materialism: The Limits of Theory in Marx," in *Fe-tishism as Cultural Discourse*, ed. Emily Apter and William Pietz, 138–139. Ithaca, N.Y.: Cornell University Press, 1993)

4. From the Portuguese:

> "Eu fui raspado para Osala em Salvador mas precisei assentar Yewa e mãe Aninha me mandou para o Rio de Janeiro porque já na época Yewa era por assim dizer um Orisa em via de extinção. Muitos já não conhe-ciam mais os oro de Yewa."

> "Eu sou de Oba, Oba quase que já morreu porque ninguém sabe assen-tar ela, ninguém sabe fazer, então eu vim para cá porque aqui eu fui raspada e a gente não vai esquecer os awo para fazer ela."

Patricia de Aquino (personal communication). I thank her for allowing me to use this information, which is from her dissertation "La construction de la personne dans le *Candomblé*" (Rio de Janeiro: National Museum, 1995). See also Patricia de Aquino and José Flavio Pessoa de Barros, "Leurs noms d'Afrique en terre d'Amérique," *Nouvelle revue d'ethnopsychiatrie* 24 (1994): 111–25. "Um Orisa em via de extinção" is an ecological expression for endangered species.

5. While rejecting naive belief in naive belief, Paul Veyne could escape this alternative only by seeing all cultures as demiurgic creators of incommensurable worlds that are unrelated to one another and to objects themselves. "It is enough to give this divine constitutive power, this power to create without the need for a preexisting model, to man's constitutive imagination" (*Did the Greeks Believe in Their Myths? An Essay on the Constitutive Imagination*, trans. Paula Wissing. Chicago: University of Chicago Press, 1988, 127). The difference between knowledge and belief, between myth and reason, is indeed eradicated, but at a price: an overall shift into creative imagination, which is unambiguously tied to the Nietzschean will to power. "They arise from the same organizing capability as the works of nature. A tree is neither true nor false; it is complicated" (123).

6. Nevertheless, the "bad faith" of the Sartrian *salaud* allows us to pass from one to the other. We shall see further on what to make of these little backroom deals.

7. "Eu sou de Dada mas como não se sabe fazer Dada, a gente entrega a Sango ou Osala par eles pegarem a cabeça da pessoa."

8. See the magnificent chapter on the sculptor's hammer in Michel Serres, *Statues: Le second livre des fondations* (Paris: Bourin, 1987), 195 ff. Referring to Michelangelo's *Pietà*, he writes: "The holes in the hands and feet of the dead Christ, the gaping wound in his flank, traces of spears or hammered nails—are they any different from the hammer wounds inflicted on the marble surface of the marble Mother by a dangerous lunatic, that Pentecost Sunday in 1972, or from the blow administered to Moses by the sculptor himself as he threw the hammer and chisel at his creature, begging it to speak? Or from the blows that carved it?" (203).

9. Emile Durkheim, *The Elementary Forms of the Religious Life*, trans. Joseph Ward Swain (New York: Free Press, 1947 [1915]).

10. Karl Marx, *Capital*, in *Karl Marx and Frederick Engels, Collected Works*, vol. 35 (New York: International Publishers, 1975 [1867]), 83.

11. See, for example: Nicholas Thomas, *Entangled Objects: Exchange, Material Culture and Colonialism in the Pacific* (Cambridge, Mass.: Harvard University Press, 1991); and especially Karl Polanyi's classic work, *The Great Transformation* (Boston: Beacon Press, 1957 [1944]).

12. In the process, I continue the slippage begun by Luc Boltanski and Laurent Thévenot in *On Justification: Economies of Worth*, trans. Catherine Porter (Princeton, N.J.: Princeton University Press, 2006), that leads from critical sociology to the sociology of criticism. One might even say that I am extending the analysis some anthropologists have carried out on the very concept of fetishism. The word brings back bad memories for anthropologists, and it does not even appear in Pierre Bonte's and Michel Izard's *Dictionnaire de l'ethnologie et de l'anthropologie* (Paris: Presses Universitaires de France, 1991). The little book by Alfonso Iacono, *Le fétichisme: Histoire d'un concept* (Paris: Presses Universitaires de France, 1992), reconstructs the history of fetishism around the notion of the refusal of the Other, and offers a detailed deconstruction of Charles de Brosses's book. Like Emily Apter's and William Pietz's book, *Fetishism as Cultural Discourse*, however, Iacono's cannot take us very far, for it never questions the virtues of anti-fetishism. While both books justifiably criticize the racist myth of a primitive religion and the systematizing follies of Auguste Comte, the authors side with Marx and Freud in all seriousness and without distancing themselves in the slightest. In their hands the social sciences, liberated from the fantasies of belief, judge Blacks and Whites alike.

13. Barbara Cassin's *L'effet sophistique* (Paris: Gallimard, 1995) is interesting for its positive portrayal of the sophists: instead of clearing their name—as scholars usually do—by making an unwarranted assumption that they were attached to pretense, Cassin argues that they never believed in belief at all. She sketches the "primitive scene" during which (for the first time?) the synonymy between what is made and what is real was broken.

14. Here I am borrowing an argument that was more subtly developed by Antoine Hennion in *La passion musicale: Une sociologie de la médiation* (Paris: Métailié, 1993), 227 ff.

15. Louis Pasteur, *Oeuvres réunies par Pasteur Vallery-Radot* (Paris: Masson, 1922) 2:13, emphasis added. For a more complete version of my analysis, see

Bruno Latour, "Les objets ont-ils une histoire? Rencontre de Pasteur et de Whitehead dans un bain d'acide lactique," L'effet Whitehead, ed. Isabelle Stengers, 197–217 (Paris: Vrin, 1994).

16. A delightful ethological description of the gestures of realism can be found in Malcolm Ashmore, Derek Edwards and Jonathan Potter, "The Bottom Line: The Rhetoric of Reality Demonstrations," Configurations 2, no. 1 (1994): 1–14.

17. I have used this trope in Bruno Latour and Steve Woolgar, Laboratory Life: The Social Construction of Scientific Facts (Beverly Hills: Sage Publications, 1979). When this book was published, the failure of social explanation was not yet apparent. I drew its conclusions only later, by removing the word "social" from the title of the second edition of the book (Laboratory Life: The Construction of Scientific Facts (Princeton, N.J.: Princeton University Press, 1986), and then by developing, with Michel Callon, the principle of generalized symmetry in Bruno Latour, The Pasteurization of France (Cambridge, Mass.: Harvard University Press, 1988), and then in Science in Action: How to Follow Scientists and Engineers through Society (Cambridge, Mass.: Harvard University Press, 1987). I had indeed detected the phenomenon, but it took me twenty years to understand the synonymy of the two verbs "to build" and "to overtake" (construire et dépasser).

18. On the history of this mastery and the notion of anthropological symmetry, see Bruno Latour, We Have Never Been Modern, trans. Catherine Porter (Cambridge, Mass.: Harvard University Press, 1993 [1991]).

19. I am not considering here the theme of verum and factum (in Vico, for example) that applies to man the old theological argument over what can be known about a world by its creator. See Amos Funkenstein, Theology and the Scientific Imagination from the Middle Ages to the Seventeenth Century (Princeton, N.J.: Princeton University Press, 1986). This theme presupposes a theology and an anthropology of the applied sciences that run totally counter to the lesson I seek to derive from fetishes.

20. The exact moment of this clinamen is not important. I tend to locate it in the exemplary anthropology of the sciences that Michel Serres has developed, from The Birth of Physics, trans. Jack Hawkes (Manchester: Clinamen Press, 2001 [1977]) to Statues, as well as in David Bloor's emblematic, Knowledge and Social Imagery, 2nd. edn. (Chicago: University of Chicago Press, 1991 [1976]). Others prefer to recognize its mark in Thomas Kuhn's The Struc-

ture of Scientific Revolutions, 3rd. edn. (Chicago: University of Chicago Press, 1996 [1962]). All that really matters here is that there is a turning point at which we see the humanities and social sciences take a new approach to the exact sciences, abandoning the four attitudes—rational reconstruction, skepticism, irrationalism, and hermeneutics—that had previously governed their relation to established knowledge. I am of course exaggerating the importance of my own discipline by asserting that its historical importance can never be overestimated. In fact, it coincided with and drew meaning and energy from the vast reversal of modernity itself—before the ecological crisis put an end to the modernist parenthesis altogether.

21. For a presentation of this failure of social explanation in the face of objects that are too difficult for it, see Michel Callon and Bruno Latour, *Les scientifiques et leurs alliés* (Paris: Pandore, 1985); and Michel Callon and Bruno Latour, eds., *La science telle qu'elle se fait: Anthologie de la sociologie des sciences de langue anglaise*, 2nd edn. (Paris: La Découverte, 1991). Provided that its consequences are understood, failure has philosophical virtues that are decidedly superior to those of success.

22. Paradoxically, far from politicizing science, science studies have allowed us to see to what extent all theories of knowledge, from the Greeks to our own day, had to bear the burden of a political definition that required separation between facts and fetishes. Freed from politics, the sciences are once again becoming fascinating, and open to anthropological description.

23. One should add to this the artifact, in the laboratory sense of a parasite mistaken for a new being. Unlike facts, artifacts are surprising, because human actions can be discovered where they have not been expected. The word thus provides a transition between the surprise of facts and the surprise of fetishes. There is no more reason for giving up the word "fetish" than for giving up the word "fact," on the pretext that the Moderns believed in belief and wanted to disqualify the former term while retaining the latter. In practice, no one has ever believed in fetishes, and everyone has always paid astute attention to facts. The two words thus remain intact. Since the phonemic difference between the French terms *fée* and *fait* ("fairy" and "fact") is not always audible, some might prefer *factiche* over *faitiche* in French. In the English "*factish*," the etymological link to "fairy" is unfortunately lost.

24. Any painter might claim that her canvas is *acheiropoeitos* (not made by human hands), but she still does not naively expect it to fall varnished from the skies — see following chapter.

25. I shall explain the meaning of this break later on. Appearances notwithstanding, technological fabrication does not escape from the either-or question; technologists are quite neatly divided between those who follow the material determinisms of *function* and those who attach themselves to the arbitrary human or social capriciousness of *form*. About this dualism, see Bruno Latour and Pierre Lemonnier, eds., *De la préhistoire aux missiles balistiques — L'intelligence sociale des techniques* (Paris: La Découverte, 1994). Also see the "*disputatio*" between the two authors in *Ethnologie française* 26:1 (1996): 17–36.

26. See Pierre Bourdieu. "La délégation et le fétichisme politique," in *Choses dites*, 185–202 (Paris: Minuit, 1987). Bourdieu gives an account of this scornful attitude toward political representation, in which anti-fetishism is pushed to the extreme: "The ministry's minister acts only provided that the minister hides his usurpation and the *imperium* he derives from it, by asserting that he is just a simple, humble minister . . . Thus the minister's symbolic violence can be practiced only with the kind of complicity granted him — through the ignorance that denegation encourages — by the very victims of that violence" (191). It would be hard to misunderstand more completely either the work of representation, or the wisdom of the represented. Only *illusio* allows sociologists not to see the glaring contradiction of anti-fetishism, whereas *illusio* is used (naively?) by the critical sociologist to describe the inability of mere actors to see the glaring contradiction of fetishism. No emperor is more exposed in his nakedness than the critical sociologist who sees himself as the only lucid person in an insane asylum.

27. Hence the fact, otherwise hard to explain, that the sociology of the "actors-themselves" can assert both that it is content with recording what the actors say, and also that it *adds* something that they never say. See Bruno Latour, *Reassembling the Social: An Introduction to Actor-Network Theory* (Oxford: Oxford University Press, 2005). Far from giving the voiceless a voice, or theorizing their simple practice, this sociology is content with foregrounding common forms of life, against the *diktats* of critical thought: whence come the notions of mediation, actor-network, translation, modes of co-

ordination, symmetry, and non-Modernity. Such infra-theoretical notions are aimed neither at expressing the practical knowledge of the actors nor at explaining it—actors do that—but only at documenting it; such a re-description of what they know might be useful to the actors. Ordinary sociologists then find themselves on the same footing as ordinary actors, just as the coastal Blacks do with the Moderns, and for the same reason.

28. Amusingly enough, the pragmatism that could be taken for the philoso-phy of practice remains so intimidated by the authoritative position of its adversaries that it is forced to depict practice as modest, limited, utilitar-ian, humanist, and convenient, thereby occupying—unquestioningly—the place critical philosophy had prepared for it. Modesty is a philosophi-cal virtue only if it decides for itself how it will keep from telling somebody what to do or from proposing foundations.

29. Art history could also offer a rich source of material for this historical an-thropology of ancient and modern iconoclasm. See the following chapter.

30. Anantha Murthy, *Bharathipura* (Madras: Macmillan, India, 1996). For a somewhat different treatment, see the last chapter of Bruno Latour, *Pan-dora's Hope: Essays on the Reality of Science Studies* (Cambridge, Mass.: Harvard University Press,1999).

31. On pariahs, see the admirable book by Viramma, *Life of an Untouchable*, eds. Josiane Racine and Jean-Luc Racine, trans. Will Hobson (London: Verso, 1997).

32. As Michel Callon and I have often demonstrated, this amounts to general-izing an ethnomethodological approach by extending it into metaphysics by way of semiotics, the only *organon* we have at our disposal that can fear-lessly maintain the diversity of modes of existence—at the price, to be sure, of putting them into language and text, a restriction that we have nonetheless sought to overcome by extending the overly-restrictive defini-tions of semiotics to things themselves. And so once again we come across the entities that had interested us from the start—under the vague head-ing of "actor-network"—and that are at once real, social, and discursive.

33. Ever since Charles de Brosses, much has been made of the material, thick, shapeless, and stupid, character of brute fetishes. But such an approach forgets that the *res extensa* is only brutish by opposition with a knowing mind. The wood, bone, clay, feather, or marble matter of these fetishes thinks, speaks, and is articulated like any other matter. A stone has noth-

ing particularly shapeless about it. Its articulations allow as much "talk-making" as those of lactic acid ferment.

34. This diagram fleshes out somewhat the overly abstract one provided in *We Have Never Been Modern*. I am replacing the double separation between nature and society on the one hand, and between purification and mediation on the other, with an object that *contains* both of them and whose presence, shape, and composition can be studied empirically.

35. On this long and tangled history, see chapter 2.

36. On these mixtures, see the quite beautiful book by Philippe Descola, *In the Society of Nature: A Native Ecology in Amazonia*, trans. Nora Scott (Cambridge: Cambridge University Press, 1994 [1986]); and Descola's literary and reflective reinterpretation in *The Spears of Twilight: Life and Death in the Amazon Jungle*, trans. Janet Lloyd (London: Flamingo, 1997).

37. Hans Jonas, in *The Imperative of Responsibility: In Search of an Ethics for the Technological Age*, trans. Hans Jonas with David Herr (Chicago: University of Chicago Press, 1984 [1979]), wanting the Moderns to follow the consequences of their actions through all their twists and turns, mistakes them for Blacks, because—without regard for the cost—he requires them to lose precisely what constitutes (what constituted) their exemplary strength: their partial irresponsibility, the rupture in the continuity of the action, the incomprehensible surprise faced with the distinct appearance of natural facts on the one hand and ethical responsibility on the other.

38. Let us not forget that we also owe another dichotomy to the Moderns and to them alone: the dichotomy between respect for ancestors on the one hand, and inventiveness free of ties to the past on the other. Reaction and revolution, tradition and innovation, emerge from the strange notion of time that is itself broken.

39. The reactionary movements of the twentieth century that tried—and are still trying—to praise paganism, and that seek to destroy the universality of reason, are horribly mistaken about what they adore and what they abhor. They depict the life of savages as desirable according to the tenets of the most prosaic exoticism, and they hate reason for what it claims to be, whereas in practice reason displays the most civilized, most refined, most socialized, most localized, and most collective form of life there is. If we have to re-anthropologize the modern world, we must do so from the top, through sciences and technologies, and not from the bottom, not

by adding credence to the way of perceiving primitives and pagans that three centuries of ordinary clericalism and racism have thought they could impose.

40. As Marshall Sahlins reminds us in *Culture in Practice* (New York: Zone Books, 2000), whereas not long ago anthropology despaired as it saw cultures coming to an end (or as it faced its own postmodern implosion), it now finds itself overwhelmed by the rebirth of new cultures that are not modern, and that *are asking* to be studied. We have yet to measure fully how much the re-equilibration in favor of Asia *lightens* Westerners: the end of Europe's guilt; the beginning of an anthropology finally as robust as the societies it must be able to analyze without destroying them.

41. In Edwin Hutchins's, *Cognition in the Wild* (Cambridge, Mass.: MIT Press, 1995), one can read—albeit in a different tradition, that of "distributed cognition"—the same externalization of the work of thought, and its "anthropologizing" in forms compatible with those of the present essay.

42. Tobie Nathan, *Principes d'ethnopsychiatrie* (Grenoble: La pensée sauvage, 1993); Tobie Nathan, *L'influence qui guérit* (Paris: Odile Jacob, 1994); and Tobie Nathan and Isabelle Stengers, *Médecins et sorciers*, in Les empêcheurs de penser en rond (Le Plessis-Robinson: Synthélabo, 1995).

43. From one year to the next, each of the canonical examples finds itself overturned by modern historiography, as in the admirable case studied by Jeffrey Burton Russell in *Inventing the Flat Earth: Columbus and Modern Historians* (New York: Praeger, 1991). How those monks have been mocked for being naive enough to believe in a flat earth in the literal sense. The author proves, with brio, that this particular belief in naive belief dates back to the nineteenth century, when there was actually nothing naive about it, since it shared in the beautiful Enlightenment scenography that was emerging from the dark ages.

44. Catherine Darbo-Peschanski's *Le discours du particulier: Essai sur l'enquête hérodotéenne* (Paris: Seuil, 1987). This has been a decisive work for me: it can serve as an overall method for bringing together the diversity of positions that has been crushed by the notion of belief. For more recent examples, see Émilie Gomart, "Methadone: Six Effects in Search of a Substance," *Social Studies of Science* 32, no. 1 (2002): 93–135.

45. We know how much trouble Galileo and his associates had trying to point to the moon with both a finger and the eyepiece of their telescope. Jean-

Marc Lévy-Leblond, "Galileo's Finger," in *Iconoclash*, ed. Bruno Latour and Peter Weibel, 146–48 (Cambridge, Mass: MIT Press, 2002).

46. Elizabeth Claverie, "La Vierge, le désordre, la critique," *Terrain* 14 (1990): 60–75; and Elizabeth Claverie, *Les Guerres de la Vierge: Une anthropologie des apparitions* (Paris: Gallimard, 2003).

47. See the models proposed by Lorraine Daston in *Biographies of Scientific Objects* (Chicago: University of Chicago Press, 2000); and Bruno Latour, *Pandora's Hope*.

48. See Pierre Lagrange, "Enquête sur les soucoupes volantes," *Terrain* 14 (1990): 76–91; and Pierre Lagrange, ed. *Ethnologie française* 23, no. 3 (1993).

49. To follow this politics of reason that saves epistemology in practice from its own theory, see Isabelle Stengers, *The Invention of Modern Science*, trans. Daniel W. Smith (Minneapolis: University of Minnesota Press, 2000 [1993]); and Barbara Cassin, *L'effet sophistique*.

50. The solution that consists in turning divinities into signifiers that are distributed by unconscious rules has allowed the structuralists to display some very nice effects of intelligibility, but we can now better measure the price they had to pay in order to elaborate this *science of nonsense*: they had to abandon the meaning of practice and deprive thought of the subtle ontology that it nevertheless asserted. The logic of the signifier is better than the delirium of "primitive thought," but the most economical solution still remains that of populating the world with the beings of whom actors speak, and according to the diverse specifications they require.

51. As I see it, this is what explains the inability of French anthropologists to understand Tobie Nathan's work; they are looking for an authenticity of ethnicity that they cannot find, and they fail to see that the originality of the Devereux Center laboratory stems precisely from its artificiality.

52. I call this work of representation that is forever in motion "transports of wills," and I see it as corresponding more closely to what is usually called the political. See Bruno Latour, "What If We Talked Politics a Little?" *Contemporary Political Theory* 2, no. 2 (2003): 143–64. Concerning those whose ancestors are monkeys, see Donna Haraway, *Primate Visions: Gender, Race and Nature in the World of Modern Science* (New York: Routledge, 1989).

53. Despite ambiguous formulations, the production of identity in the form of therapy renewed by Tobie Nathan is not based on culturalism at all, but

on the voluntaristic and sometimes violent creation of an affiliation that is exactly as artificial as the framework of the session itself. For a recent description of this, see Tobie Nathan, "La haine: Réflexions ethnopsychanalytiques sur l'appartenance culturelle," *Nouvelle revue d'ethnopsychiatrie* 28 (1995): 7–17. This is an essential point, for it is what distinguishes ethnopsychiatry from a reactionary way of thinking that, on the contrary, pretends to enclose an ethnic identity once and for all within a natural affiliation. Here, too, artifice is the friend and not the enemy of reality, whether in terms of the laboratory setup or the creation of affiliations.

54. See the text by Isabelle Stengers that constitutes the second half of Nathan and Stengers, *Médecins et sorciers*. The "will to do science" deprives the charlatan-turned-researcher of the ability to understand the influence he is exercising: see Isabelle Stengers, *La volonté de faire science: A propos de la psychanalyse* (Paris: Les Empêcheurs de penser en rond, 1992). This book allows for a positive and non-critical reading of Mikkel Borch-Jacobsen's *Remembering Anna O.: A Century of Mystification*, trans. Kirby Olson with Xavière Callahan and Mikkel Borch-Jacobsen (New York: Routledge, 1996 [1995]). According to Stengers, in applying to humans an epistemological model that no researcher had ever applied to objects, and thus imitating a nonexistent scientific model, psychiatrists have failed to see the specific originality of the therapy. Paradoxically, one must treat humans as Pasteur treats his lactic acid ferment in order to begin to "make them talk" in an interesting way. On this whole confusion among theories of agency, see Isabelle Stengers, *La vie et l'artifice: Visages de l'émergence* (Paris: La Découverte; Le Plessis-Robinson: Les Empêcheurs de penser en rond, 1997).

55. Lévi-Strauss shaped the literary genre of this theory of science long ago, but the theory is found again in a fascinating article by Marika Moisseeff, "Les objets culturels aborigènes ou comment représenter l'irreprésentable," *Genèses: Sciences sociales et histoire* 17 (1994): 8–32. Moisseeff claims to be giving up linguistic metaphors and returning to objects, however. The most dignified state to which she can compare an object is that of "pure signifier" (28). In Marc Augé's *Le dieu objet* (Paris: Flammarion, 1988), one can likewise find no higher way of speaking of god-objects than to turn them into thoughts: "Telling needs matter at once to represent itself, to speak itself, and to actualize itself; and matter needs telling to become an object of thought" (140). The degree to which ethnologists' positivism,

concerning their access to nature, shapes their definition of culture in advance is never adequately measured. It is very difficult to find ethnographies that have been able to rid themselves of Kant.

56. We must not be surprised that Judaism, Christianity, and Islam have regularly condemned the mobilization of divinities through witchcraft, but have all allowed therapies—under different forms—to proliferate without being able to integrate them into their theology. On Judaism's "misunderstanding" regarding the fight against idols, see Moshe Halbertal and Avishai Margalit, *Idolatry* (Cambridge, Mass.: Harvard University Press, 1992).

57. I use the expression "person formation" to designate this particular mediation, as different from the one studied here as it is from the "will formation" by which identities and representations are fabricated. See chapter 3 and my "How to Be Iconophilic in Art, Science, and Religion?" in *Picturing Science, Producing Art*, eds. Caroline A. Jones and Peter Galison, 418–40 (New York: Routledge, 1998). It is clear that the difference between the substance-less gods invoked here, and those of rationalist theology, is just as great as the difference between the objects of science and the dreams of epistemologists, or between divinities and mysterious spirits or supernatural beings.

58. That is the difference between the work Michel Callon and I initiated fifteen years ago on "actor-networks," and the project we have begun only recently. We seek to replace the black and white television image of the "actor-networks" with a color picture, by fabricating enough analyzers to register the principal contrasts that seem important to the "actors-themselves," our only masters.

59. See Jeanne Favret-Saadra's classic book *Deadly Words: Witchcraft in the Bocage*, trans. Catherine Cullen (Cambridge: Cambridge University Press, 1980 [1977]). See also Jeanne Favret-Saadra and Josée Contreras, "Ah! La féline, la sale voisine . . ." *Terrain* 14 (1990): 20–31.

60. Shirley Strum, *Almost Human: A Journey into the World of Baboons* (New York: Random House, 1987). See also Frans de Waal, *Peacemaking among Primates* (Cambridge, Mass.: Harvard University Press, 1989).

61. Rainer, Maria Rilke "Eighth Letter," in *Letters to a Young Poet*, trans. Joan M. Burnham (Novato, Calif.: New World Library, 2000 [1903]), 80.

62. Nathan, *L'Influence qui guérit*. Nathan contrasts "anguish" and "fright." An-

guish is supposed to have no outside cause and comes entirely from the
subject's own projection; fright, on the contrary is clearly the realization
that there is something outside that has caused you to be frightened.

63. Free association only retains in language the faraway echo of the ontological substitutions that are themselves a means, among many others, of
exploring being-as-Other. This is another version of the link that "makes
talk" in the artificial laboratory of the couch.

64. It is not as dangerous as one might think to borrow the old language of
psychoanalysis, for in the long run, once the couch, the doctor's office, the
money, the professional associations, the controversies, the works, the
style, and the ancestors have been reintroduced, we end up with a setup
that is just as artificial, just as unpsychological, and thus just as *interesting* as that of ethnopsychiatry. One advantage of symmetry is that all the
"head-shrinkers" can study themselves the same way. Ethnopsychiatry re-
stores its *material and collective culture* to a psychoanalytical practice whose
theory claimed to have to do with subjects, and to be a science thirsting
for truth. Let us note that simulation, like *factishes*, refuses to respect the
either-or choice: is it real, or is it simulated? In the laboratory as on the
couch, simulation refuses that very alternative, and opens up another path
between artificiality and truth. See Borch-Jacobsen, *Remembering Anna O*.

65. Gilles Deleuze and Félix Guattari, *Anti-Oedipus: Capitalism and Schizophrenia*,
trans. Robert Hurley, Mark Seem, and Helen R. Lane (Minneapolis: University of Minnesota Press, 1983 [1972]).

66. This is the case in particular with *interaction transformations*, which are
called, rather too hastily, "techniques." On this very particular form of
mediation, see Bruno Latour, "On Interobjectivity—with discussion by
Marc Berg, Michael Lynch and Yrjo Engelström," *Mind Culture and Activity*
3, no. 4 (1996): 228–45. See also *Pandora's Hope*, chapter 6.

67. In my jargon, these are "immobile mobiles." See *Pandora's Hope*, chapter 2.

68. Let us not forget that the very idea of an ineffable practice came simply
from epistemological illusions about the explicit formalism of scientific
speech. My colleagues and I have learned at our own expense how hard it
is to express the work of the sciences *in words*. But by this very token, no
practice is easier or more difficult to make explicit than any other. On the
work of formalism, see the essay, the title of which alone is a whole program, by Brian Rotman, *Ad Infinitum: The Ghost in Türing's Machine. Taking God*

out of Mathematics and Putting the Body Back In: An Essay in Corporeal Mathematics (Stanford, Calif.: Stanford University Press, 1993).

69. If we could not succeed in answering these questions, the symmetry would be broken and Whites would indeed remain without *factishes*. But if we begin to search empirically for doubly broken and then cleverly mended *factishes*, we should be able to find them, as for example in the astonishing roles played by medications. See Andrew Lakoff, *Pharmaceutical Reason: Knowledge and Value in Global Psychiatry* (Cambridge: Cambridge University Press, 2006), and for drugs, Émilie Gomart, "Methadone." This is the new direction for sociologists of medicine. For the state of the art, see Anne-Marie Mol and Marc Berg, eds., *Differences in Medicine: Unraveling Practices, Techniques and Bodies* (Durham, N.C.: Duke University Press, 1998).

70. This section is adapted from "Factures/Fractures: From the Concept of Network to that of Attachment," trans. Monique Girard Stark, *Res 36* (Autumn, 1999): 20–31.

71. Émile Benveniste, "The Active and Middle Voice in the Verb," in *Problems in General Linguistics*, 145–71 (Coral Gables, Fla: University of Miami Press, 1971). The expression of "middle" is a later rationalization, once the active and passive had become evident. In this brief and critical chapter, Benveniste cast the "middle" as ancestral to the passive form; this more ancient opposition distinguishes it from the active: "One can diversify at will the play of these oppositions . . . they always finally come down to situating positions of the subject with respect to the process, according to whether it is exterior or interior to it, and to qualifying it as agent, depending on whether it effects, in the active, or whether it effects while being affected, in the middle" (149–50).

72. See Pierre Legendre, *Sur La Question Dogmatique En Occident* (Paris: Fayard, 1999). The new influence by Pierre Legendre can be explained, from my perspective, by this reversal: suddenly before our eyes—sometimes in our own children—we have these emancipated beings, a state only aspired to—or feared—by preceding generations, who would never have become detached, so firmly did the chains of the past hold them. The experiment is now complete: as Legendre affirms with prophetic violence, "You the fathers, you have given birth to the living-dead." His solution, derived more from Lacan than from roman law, unfortunately amounts to forgoing attachments in order to impose upon subjects a sovereign power

defined by void alone, making disappear even more radically the multiple sources of *faire-faires*.

73. The concept of "affordance" is so powerful because it permits the deployment of the middle voice in psychology. See Bruno Latour, *Reassembling the Social*. See also the work of Laurent Thévenot on the forms of ordinary action: Laurent Thévenot, *L'action au pluriel. Sociologie des régimes d'engagement* (Paris: La Découverte, 2006).

74. The question concerning the attachment of properties is not any easier to resolve than that of divinities, and the key concept of externality does not suffice to end the discussion, despite its pretensions of achieving closure. See Michel Callon, ed., *The Laws of the Markets* (Oxford: Blackwell, 1998). The arguments on the freedom of choice or the organization of the market rehearse exactly the same theories of inaction as those of the social sciences.

75. In addition to the summit reached by Pietz in "Fetishism and Materialism: The Limits of Theory in Marx," see the dizzying analysis of Simon Schaffer, on the relation between history and scientific assessment of gold and the accusation of fetishism: Simon Schaffer, "Golden Means: Assay Instruments and the Geography of Precision in the Guinea Trade," in *Instruments, Travel and Science: Itineraries of Precision from the Seventeenth to the Twentieth Centuries*, ed. Marie-Noelle Bourguet, Christian Licoppe and Otto Sibum (London: Harwood Academic Press, 2001).

76. As much could be said about an "internal" exoticism (invented by critical theory, in particular the Frankfurt School), which has transformed all European and American cultures into a manipulated mass, also bizarrely attached. Critical theory plays for the center the same role of exoticizing alterity as the conceptualization of the fetish performs for the periphery.

77. On this issue of a great divide, see Bruno Latour, *We Have Never Been Modern*. See also the important work accomplished by anthropologists on revising the categories of culture, once one removes the obstacle posed by an opposing category of nature: Philippe Descola, *Par delà nature et culture* (Paris: Gallimard, 2005).

78. Isabelle Stengers, *Cosmopolitiques* (Paris: La découverte, 1996).

79. This is the objective of the effort undertaken in Bruno Latour, *Politics of Nature: How to Bring the Sciences into Democracy*. trans. Catherine Porter (Cambridge, Mass: Harvard University Press, 2004). The definition of a collective capable of assembling a common world without having recourse to

the two traditional compendiums of Nature and Society, bicameralism is ill adapted to the contemporary situation.

80. A different theory of action, focused on China, can be found in François Jullien, *The Propensity of Things: Toward a History of Efficacy in China*, trans. Janet Lloyd (New York: Zone Books, 1995). This one does not coincide very well with the Westerners' theory either, for it ignores both immanence and transcendence, subjects and objects. According to Jullien's interpretation, the Chinese seem to supply practice with a language that the Whites have never abandoned, but that their philosophy, for interesting political reasons, has often sought to disown.

81. And let us not forget the etymology that will remind us, quite appropriately, that the French word *marionnette* is a term of endearment for "little Saint Marys," intercessory virgins *par excellence*.

82. This is why there is no way to criticize constructivism without criticizing the theory of action implicit in the notion, as Ian Hacking made painfully clear in *The Social Construction of What?* (Cambridge, Mass: Harvard University Press, 1999).

83. Discoveries in theology are rare; but Whitehead's discovery of the creature god certainly counts as one. Indeed, for Whitehead it was less a matter of discovering than of understanding, through a different language, what everybody had already understood in a different way: Whitehead's god is *incarnate*. "All actual entities share with God this characteristic of self-causation. For this reason every actual entity also shares with God the characteristic of transcending all other actual entities, including God" (*Process and Reality: An Essay in Cosmology*. New York: Free Press, 1978, 222). To believe that God will now submerge himself among his creatures is to keep on repeating the same mistake. Creatures are not immanent. Mediations, events, passages, and *factishes* can be used neither for drowning, nor for dissolving, but only for producing. They come to pass. They differ.

What Is Iconoclash?

Adapted from "What Is Iconoclash? Or Is There a World Beyond the Image Wars?" Introduction to *Iconoclash, Beyond the Image-Wars in Science, Religion and Art*, ed. Peter Weibel and Bruno Latour, 14–37 (Cambridge, Mass.: MIT Press, 2002).

1. "Freud concentrates all the counter-religious force of Biblical monotheism in Akhenaten's revolution from above. This was the origin of it
   all. Freud stresses (quite correctly) the fact that he is dealing with the
   absolutely first monotheistic, counter-religious, and exclusivistically in-
   tolerant movement of this sort in human history. The similarity of this
   interpretation to Manetho's is evident. It is this hatred brought about by
   Akhenaten's revolution that informs the Judeophobic texts of antiquity."
   (Jan Assmann, *Moses the Egyptian: The Memory of Egypt in Western Monotheism*,
   Cambridge, Mass.: Harvard University Press, 1997), 167.

2. On the genealogy of fanatics and other *Schwärmer*, see the fascinating account of Dominique Colas, *Le glaive et le fléau: généalogie du fanatisme et de la
   société civile* (Paris: Grasset,1992). See also Olivier Christin, *Une révolution
   symbolique* (Paris: Minuit, 1991).

3. See Marie-José Mondzain, *Image, icône, économie. Les sources byzantines de l'ima-
   ginaire contemporain* (Paris: Le Seuil, 1996), 120.

4. See Han Belting, "Beyond Iconoclasm. Nam June Paik, the Zen Gaze and
   Escape from Representation," in *Iconoclash: Beyond the Image-Wars in Science,
   Religion and Art*, ed. Peter Weibel and Bruno Latour, 390–411 (Cambridge,
   Mass.: MIT Press, 2002); and Richard Powers, "The Artist's Bedlam," in
   Iconoclash, 476–78.

5. See Heather Stoddard, "The Religion of Golden Idols," in *Iconoclash*, 436–
   55.

6. For instance the *Iconoclasme* exhibition in Bern and Strasbourg in 2001.
   The Bern exhibition was entirely built in the honor of the courageous icon
   breakers who had freed the city from the power of image, leading to the
   superior symbolism of the cross all the way to a diorama where wax figures
   were melting useless calices and reliquaries to mould useful Swiss gold
   coins. But in a nice *iconoclash* the last room showed the permanent rem-
   nants of the broken statues, which had been transmogrified from hideous
   idols into art work piously conserved. No indication was given the visitors of any possible *iconoclash*. See the exhibition catalogue, *Iconoclasme.
   Vie et mort de l'image médiévale*, ed. Cécile Dupeux, Peter Jezler, et al. (Paris:
   Somogy editions d'art, 2001). The same iconoclastic piety can be seen in
   the Louvre exhibition by Régis Michel, *La peinture comme crime*, Paris: Ré-
   unions des musées nationaux, 2002.

7. Miguel Tamen. *Friends of Interpretable Objects* (Cambridge, Mass.: Harvard
   University Press, 2001).

8. See Joseph Koerner, "The Icon as Iconoclash," in *Iconoclash*, 164–213; and Marie-José Mondzain, "The Holy Shroud: How Invisible Hands Weave the Undecidable," in *Iconoclash*, 324–35.

9. See Jean-Marc Lévy-Leblond, "Galileo's Finger," in *Iconoclash*, 146–48.

10. See Lorraine Daston, "Nature Paints," in *Iconoclash*, 136–38; see also Lorraine Daston and Peter Galison, "The Image of Objectivity," *Representation* 40 (2001): 81–128; and Peter Galison, "Judgment against Objectivity," in *Picturing Science, Producing Art*, ed. Caroline A. Jones and Peter Galison, 327–59 (New York: Routledge, 1998).

11. Michael Taussig, *Defacement: Public Secrecy and the Labor of the Negative* (Stanford, Calif.: Stanford University Press, 1999).

12. See Ramon Sarrô, "The Iconoclastic Meal: Destroying Objects and Eating Secrets Among the Baga of Guinea," in *Iconoclash*, 227–30; and Patricia de Aquino, "No Freeze-Frame on God," in *Iconoclash*, 234–35.

13. See Pierre Centlivres, "Life, Death and Eternity of the Buddhas in Afghanistan," in *Iconoclash*, 75–77; Jean-Michel Frodon, "The War of Images, or the Bamiyan Paradox," in *Iconoclash*, 221–23; and Jean-François Clément, "The Empty Niche of the Bamiyan Buddha," in *Iconoclash*, 218–20. See also Pierre Centlivres, *Les Boudhas d'Afghanistan* (Lausanne: Favre, 2001).

14. Assmann, *Moses the Egyptian*. See also William Pietz, "The Sin of Saul," in *Iconoclash*, 63–65; Raymond Corbey, "Image-breaking on the Christian Frontier," in *Iconoclash*, 69–71; and Anne-Christine Taylor, "The Face of Indian Souls: a Problem of Conversion," in *Iconoclash*, 462–64.

15. See Tobie Nathan, "Breaking Idols . . . a Genuine Request for Initiation," in *Iconoclash*, 470–74; and Robert Koch, "The Critical Gesture in Philosophy," in *Iconoclash*, 524–36.

16. "Either those statues are linked to idolatrous beliefs," commented the Mullah, "or they are mere stones; in the first case, Islam requests them to be destroyed; in the second case, why bother if they are destroyed?" Pierre Centlivres, *Saints, sainteté et martyr: La fabrication de l'exemplarité. Actes du colloque de Neuchâtel, 27–28 novembre 1997*, 141. Paris, 2001.

17. See Kroener, "The Icon as Iconoclash"; and Olivier Christin, "The Idol King?," in *Iconoclash*, 65–68.

18. See Peter Galison, "Images Scatter into Data, Data Gather into Images," in *Iconoclash*, 300–323; Christian Kassung and Thomas Macho, "Imaging Processes in Nineteenth Century Medicine and Science," in *Iconoclash*, 336–47; Jörg Huber, "On the Credibility of World-Pictures," in *Iconoclash*,

520–23; and Hans-Jörg Rheinberger, "Auto-Radio-Graphics," in *Iconoclash*, 516–19.

19. See Simon Schaffer, "The Device of Iconoclasm," in *Iconoclash*, 498–515.

20. See Hans Belting, "Beyond Iconoclasm," in *Iconoclash*, 390–411; Boris Groys, "Iconoclasm as an Artistic Device: Iconoclash Strategies in Film," in *Iconoclash*, 282–95; and Peter Weibel, "Is Modern Art Iconoclast? Up To and Beyond the Crisis of Representation," in *Iconoclash*, 570–670.

21. See Dario Gamboni, "Image to Destroy, Indestructible Image," in *Iconoclash*, 88–135; and Nathalie Heinich, "Baquié at Malpassé: An 'Adventure' in Contemporary Iconoclasm?" in *Iconoclash*, 417–20.

22. See Albena Yaneva, "Challenging the Visitor to Get the Image: On the Impossible Encounter of a Pig and an Adult," in *Iconoclash*, 421–22; and Adam Lowe, "To See the World in a Square of Black," in *Iconoclash*, 544–67.

23. See for instance the magnificent Tim J. Clark, *Farewell to an Idea: Episodes from a History of Modernism* (New Haven, Conn.: Yale University Press, 1999).

24. See Peter Sloterdijk, "Analytical Terror," in *Iconoclash*, 352–59; and Weibel, "Is Modern Art Iconoclast?"

25. See Hans-Ulrich Obrist, "Milano Triennale 68: A Case Study and Beyond," in *Iconoclash*, 360–83; John Tresch, "Did Francis Bacon Eat Pork? A Note on the Tabernacle in 'New Atlantis,'" in *Iconoclash*, 231–33; and Lowe, "To see the World in a Square of Black."

26. See Gamboni, "Image to Destroy."

27. See Kroener, "The Icon as Iconoclash"; and Mondzain, "The Holy Shroud."

28. See Galison, "Images Scatter into Data"; and Schaffer, "The Device of Iconoclasm."

29. See Weibel, "Is Modern Art Iconoclast?"

30. See Kroener, "The Icon as Iconoclash."

31. See chapter one.

32. See the striking case of La Fontaine, in fig. 1. For another interpretation, see Gamboni, "Image to Destroy."

33. On the striking case of Francis Bacon, see Tresch, "Did Francis Bacon Eat Pork?" On the sad case of the Jerusalem temple, see Moshe Halbertal, "God Does Not Live There Anymore," in *Iconoclash*, 60–62.

34. See Jean-François Clément, "L'image dans le monde arabe: interdits et possibilités," in *L'image dans le monde arabe*, ed. G. Beaugé, J.-F. Clément, 11–42 (Paris: Editions du CNRS, 1995).

35. Sigmund Freud, *L'homme Moïse et la religion monothéiste. Trois essays* (Paris: Gallimard, 1996).

36. The difference between the two types of murders might explain some of the strange visual features of Freud's cabinet. See Lydia Marinelli, "Freud's Fading Gods," in *Iconoclash*, 468–69; and more largely what Andreas Mayer calls "psychic objects" in "The Fetish-Scientist, or Why Would Anyone Force Anyone to Kiss the Bust of Franz Josef Gall?" in *Iconoclash*, 465–67.

37. Midrash Rabbah, *Noah* 38: 13. Kindly translated for me by Shai Lavi.

38. See Nathan, "Breaking Idols."

39. Nowhere is it clearer than in science studies, my original field, where it organizes every position between realism and constructivism. See Ian Hacking, *The Social Construction of What?* (Cambridge, Mass.: Harvard University Press, 1999).

40. Anantha Murthy, *Bharathipura* (Madras: Macmillan, India, 1996); see chapter 1.

41. See Koch, "The Critical Gesture in Philosophy"; and Peter Sloterdijk, *Critique of Cynical Reason*, trans. Michael Eldred (Minneapolis: University of Minnesota Press, 1987).

42. See Schaffer, "The Device of Iconoclasm."

43. As recalled in Centlivres, "Life, Death and Eternity of the Buddhas in Afghanistan," Mullah Omar made a sacrifice of 100 cows, a very costly hecatomb by Afghan standards, as atonement for having *failed* to destroy the Buddhas for so long: 100 cows to ask remission for this horrible sin of eleven centuries, without wrecking them.

44. See Galison, "Images Scatter into Data."

45. See Stoddard, "The Religion of Golden Idols."

46. See Schaffer, "The Device of Iconoclasm."

47. See Mondzain, "The Holy Shroud."

48. Denis Laborde, "Vous avez-tous entendu son blasphème? Qu'en pensez-vous? Dire la Passion selon St Matthieu selon Bach," *Ethnologie française* 22 (1992), 320–33.

49. Boris Groys, *The Total Art of Stalinism: Avant-Garde, Aesthetic Dictatorship, and Beyond*, trans. Charles Rougle (Princeton, N.J.: Princeton University Press, 1992).

50. Heather Stoddard, *Le Mendiant de l'Amdo* (Paris: Société d'ethnographie, 1985).

51. Bruno Pinchard, "Tender Blasphemy: Three States of the Image, Three States of Love in the Renaissance," in *Iconoclash*, 151–54.

52. See Dominique Linhardt, "All Windows Were Open, But Nothing Happened? Nothing! Well . . . Except a Lot!" in *Iconoclash*, 148–50; and Sloterdijk, "Analytical Terror."

53. See Michael Taussig, "Old Glory," in *Iconoclash*, 82–83.

54. Political correctness is part of this attitude: scouting everywhere for good occasions to be scandalized.

55. See Koch, "The Critical Gesture in Philosophy."

56. For the mechanism of scandal mongering in contemporary art, see Heinich, "Baquié at Malpassé"; and Dario Gamboni, *The Destruction of Art. Iconoclasm and Vandalism since the French Revolution* (London: Reaktion Books, 1996). The typical mechanism for seeing objects as tokens has been proposed by René Girard, *Things Hidden Since the Foundation of the World* (Stanford, Calif.: Stanford University Press, 1987).

57. Louis Réau, *Histoire du vandalisme. Les monuments détruits de l'art français*, Edition augmentée par Michel Fleury et Guy-Michel Leproux (Paris: Bouquins, 1994). See also André Chastel, *The Sack of Rome—1527* (Princeton, N.J.: Princeton University Press, 1983).

58. Censorship may be one aspect of the Ds: tearing down or hiding images for the sake of protecting other images and choosing the wrong target.

59. Lowe, "To See the World in a Square of Black."

60. See Obrist, "Milano Triennale 68"; and Peter Geimer, "Searching for Something: On Photographic Revelation," in *Iconoclash*, 143–45.

61. See Brigitte Derlon, "From New Ireland to a Museum: Opposing Views of the Malanggan," in *Iconoclash*, 139–42; and Sarrô "The Iconoclastic Meal."

62. See Z. S. Strother, "Iconoclasm by Proxy," in *Iconoclash*, 458–59. Other cases could be found of retrospective destruction in technology: asbestos used to be the "magic material" before its producers were accused of killing thousands of people with it; DDT used to be the magic pesticide before being accused of the same crimes. For an account of this retrospective accusation around the notion of "after-effect," see Ulrich Beck, *Ecological Politics in an Age of Risk* (Cambridge: Polity Press, 1995).

63. Bruno Pinchard, "On a Suspended Iconoclastic gesture," in *Iconoclash*, 456–57.

64. Sloterdijk, *Critique of Cynical Reason*.

65. Christian Boltanski, personal communication.

66. I proposed a test to Maurizio Cattelan: to replace the Pope (whom everyone, but perhaps not the Poles, expects to see smashed to the ground) by someone whose destruction would really trigger the intellectuals' indignation: for instance to show Salman Rushdie shot to death by an Islamist bullet. . . . Too horrifying, too scandalous, I was told (Hans-Ulrich Obrist, personal communication). Ah, so the Pope can be struck but not someone *really* worthy of respect in the eyes of the critically minded! But when I proposed what appeared to be a true sacrilege and not a cheap one, what was I after? Another provocation directed at faithful critics instead of faithful Papists? Who is to tell? I can't even be sure I understand the reactions of those who recoiled in horror at my suggestion.

67. Tobie Nathan. *L'influence qui guérit* (Paris: Odile Jacob, 1994), and chapter 1 of the present book.

68. Jean Wirth, "Faut-il adorer les images? La théorie du culte des images jusqu'au concile de Trente," in *Iconoclasme. Vie et mort de l'image médiévale*, ed. Cécile Dupeux, Peter Jezler, and Jean Wirth, 28–37 (Somogy éditions d'art, 2001). In his nice visual summary of images and their prototype, Wirth manifests once more the perfect contradiction of the argument since in order to show the difference between respect for the image (*dulie*) and adoration of the model (*latrie*), he is forced, by necessity, to draw two images—one for the prototype and another one for the original.

69. See Kroener, "The Icon as Iconoclash."

70. See Mondzain, "The Holy Shroud"; and Stoddard, "The Religion of Golden Idols"; and Joseph Koerner's beautiful essay, "Hieronymus Bosch's World Picture," in *Picturing Science, Producing Art*, ed. Caroline A. Jones and Peter Galison, 297–326 (New York: Routledge, 1998). See also the notion of "dissimiles" in Georges Didi-Huberman, *Fra Angelico. Disssemblance et figuration* (Paris: Flammarion, 1990).

71. Louis Marin, *Opacité de la peinture. Essais sur la representation* (Paris: Usher, 1989).

72. The word "cascade" to describe this succession was first used by Trevor Pinch in "Observer la nature ou observer les instruments," *Culture technique* 14 (1985): 88–107. See also Mike Lynch and Steve Woolgar, eds. *Representation in Scientific Practice* (Cambridge, Mass.: MIT Press, 1990); and Caroline A. Jones and Peter Galison, eds. *Picturing Science, Producing Art*.

73. For a description of this cascading effect see Bruno Latour, *Pandora's Hope: Essays on the Reality of Science Studies* (Cambridge, Mass.: Harvard University Press, 1999), chapter 2.

74. See Galison, "Images Scatter into Data."

75. See Rheinberger, "Auto-Radio-Graphics."

76. See Huber, "On the Credibility of World-Pictures"; and Kassung and Macho, "Imaging Processes in the Nineteenth Century Medicine and Science."

77. This is why it took so long for the scientific gaze to accommodate their sight to those strange new scientific images, as is magnificently shown in Lorraine Daston and Katharine Park, *Wonders and the Order of Nature* (Cambridge: Zone Books, 1999).

78. Peter Galison, *Image and Logic: A Material Culture of Microphysics* (Chicago: University of Chicago Press, 1997).

79. See Caroline Jones, "Making Abstraction," in *Iconoclash*, 412–16; Belting, "Beyond Iconoclasm"; and Weibel, "Is Modern Art Iconoclast?"

80. For instance, George Steiner, *Real Presences* (Chicago: University of Chicago Press, 1991). Also, Jean Clair, *Considérations sur l'état des beaux arts. Critique de la modernité* (Paris: Gallimard, 1983). For a file about the debate around contemporary arts see P. Barrer, *(Tout) l'art contemporain est-il nul? Le débat sur l'art contemporain en France avec ceux qui l'ont lancé. Bilan et perspective* (Lausanne: Favre, 2000).

81. James Elkins, *Why are our Pictures Puzzles?* (London: Routledge, 1999). It could even be argued that it is from looking at (probably Dutch) paintings that philosophers of science have taken their ideas of the visible world and their model-copy epistemology. See the classic, Svetlana Alpers, *The Art of Describing* (Chicago: University of Chicago Press, 1983).

82. See Lowe, "To see the World in a Square of Black"; Yaneva, "Challenging the Visitor to Get the Image"; and Michel Jaffrennou, "Ceci n'est plus une image! This is not a Picture!" in *Iconoclash*, 479–82.

"Thou Shall Not Freeze-Frame"

Adapted from "'Thou Shall Not Freeze-Frame,' or How Not to Misunderstand the Science and Religion Debate," in James D. Proctor, ed., *Science, Religion, and the Human Experience*, 27–48 (Oxford: Oxford University Press, 2005).

1. William James. *The Varieties of Religious Experience: A Study in Human Nature* (Harmondsworth: Penguin, 1986 [1902]).

2. Bruno Latour, "Thou Shall Not Take the Lord's Name in Vain — Being a Sort of Sermon on the Hesitations of Religious Speech," *Res* 39 (Spring 2002): 215–35. For a general inquiry into the background of science and religion comparison see chapter 2.

3. Their specifications are thus different from those of divinities we studied earlier, chapter 1.

4. See chapter 1 for the emergence of belief in the encounter with anti-fetishism.

5. Ludwig Wittgenstein, "Lectures on Religious Beliefs," in *Lectures and Conversations on Aesthetics, Psychology and Religious Belief*, ed. Cyril Barrett (Oxford: Basil Blackwell, 1966), 58.

6. See Pascal Boyer, *Religion explained: The Evolutionary Origins of Religious Thought* (New York: Basic Books, 2002). Evolutionary theology shares with the old natural theology of the eighteenth century the admiration for the marvelous adjustment of the world. It does not matter much if this leads to an admiration for the wisdom of God, or of Evolution, since in both cases it is the marvelous fit that generates the impression of providing an explanation. Darwin would destroy the natural theology of old as well as this other natural theology based on evolution: there is no fit, no sublime adaptation, no marvelous adjustment. But the new natural theologians have not realized that Darwin dismantled their church even more quickly than that of those of their predecessors, whom they despise so much.

7. Alfred North Whitehead, *Religion in the Making* (New York: Fordham University Press, 1926).

8. Under William James's pen, science is a "she" — a nice proof of political correctness before its time.

9. For a much more advanced argument about visualization in science, see Peter Galison, *Image and Logic* (Chicago: University of Chicago Press, 1997); and Carrie Jones and Peter Galison, eds. *Picturing Science, Producing Art* (New York: Routledge, 1998). See also chapter 2 on the notion of cascade.

10. For all of what follows, see chapter 2.

11. See Joseph Leo Koerner, *The Reformation of the Image* (Chicago: University of Chicago Press, 2004); and chapter 2.

12. Louis Marin, *Opacité de la peinture. Essais sur la representation* (Paris: Usher, 1989).

Index

Bruno Latour is a professor at Sciences Po, Paris. Among his many books are *Reassembling the Social: An Introduction to Actor-Network Theory* (2005); *Politics of Nature: How to Bring the Sciences into Democracy* (2004); *Pandora's Hope: Essays on the Reality of Science Studies* (1999); *Aramis, Or, The Love of Technology* (1996); *We Have Never Been Modern* (1993); *The Pasteurization of France* (1988); *Science In Action: How to Follow Scientists and Engineers through Society* (1987); and with Steve Woolgar, *Laboratory Life: The Social Construction of Scientific Facts* (1979).

Library of Congress Cataloging-in-Publication Data
Latour, Bruno.
On the modern cult of the factish gods / Bruno Latour ;
first chapter translated by Catherine Porter and Heather MacLean.
p. cm. — (Science and cultural theory)
Includes bibliographical references and index.
ISBN 978-0-8223-4816-0 (cloth : alk. paper)
ISBN 978-0-8223-4825-2 (pbk. : alk. paper)
1. Religion and science.   2. Cultural psychiatry.
3. Fetishism.   4. Iconoclasm.
I. Latour, Bruno. Petite réflexion sur le culte
moderne des dieux faitiches. English.   II. Title.
III. Series: Science and cultural theory.
BL240.3.L38 2010
201'.65 — dc22   2010024239